Building Soils *for* Better Crops

2ND EDITION

Fred Magdoff and Harold van Es

SAN, the national outreach partner of USDA's SARE program, is a consortium of individuals, universities and government, business and nonprofit organizations dedicated to the exchange of information on sustainable agricultural systems.

For more information about the Sustainable Agriculture Network or its other publications see www.sare.org or contact:

Andy Clark
SAN Coordinator
c/o AFSIC, Room 304
National Agricultural Library
10301 Baltimore Ave.
Beltsville, MD 20705-2351
(301) 504-6425; (301) 504-6409 (fax)
san@nal.usda.gov

SARE is a competitive grants program that provides funding for research and education projects that promote agricultural systems that are profitable, environmentally sound and enhance the viability of rural communities nationwide.

For more information about the SARE program and SARE grants contact:
Office of Sustainable Agriculture Programs
U.S. Department of Agriculture
1400 Independence Ave., S.W., Stop 2223
Washington, D.C. 20250-2223
(301) 405-3186; (301) 314-7373 (fax)
vberton@wam.umd.edu

To order copies of this book, send a check or purchase order for $19.95 + $3.95 s/h to:

Sustainable Agriculture Publications
Hills Building, Room 10
University of Vermont
Burlington VT 05405-0082

To order SAN publications by credit card (VISA, MC), please call (802) 656-0484.

For first book sent outside of North America, please add $6. Add $2.50 for each additional book. Please include your mailing address and telephone number.

Previous titles in SAN's handbook series include 1: *Managing Cover Crops Profitably, 1st Edition,* edited by the staff of Rodale Institute; 2: *Steel in the Field: A Farmer's Guide to Weed Management Tools,* 1997, edited by Greg Bowman; 3. *Managing Cover Crops Profitably, 2nd Edition* written by Greg Bowman, Craig Cramer and Chris Shirley and edited by Andy Clark.

Library of Congress Cataloging-in-Publication Data

Magdoff, Fred, 1942-
   Building soils for better crops / by Fred Magdoff and Harold van Es.--2nd ed.
     p. cm. -- (Sustainable Agriculture Network handbook series ; bk. 4)
   Includes bibliographical references.
   ISBN 1-888626-05-4
     1. Humus. 2. Soil management. I. Van Es, Harold, 1958- II. Sustainable Agriculture Network. III. Title. IV. Series.

S592.8 .M34 2000
631.4'17--dc21
                       00-029695

Graphic design, interior layout and cover design by Andrea Gray. Cover illustration by Frank Fretz. Some interior illustrations by Bonnie Acker and Elayne Sears. Copyediting by Tawna Mertz, Valerie Berton and Andy Clark. Indexing by Peggy Holloway. Printing by Jarboe Printing, Washington, DC.

Contents

Preface   *v*

Introduction   *vii*

PART ONE

THE BASICS OF SOIL ORGANIC MATTER, PHYSICAL PROPERTIES, AND NUTRIENTS

1 Healthy Soils   3

2 What is Soil Organic Matter?   9

3 The Living Soil   13

4 Why is Organic Matter So Important?   21

5 Amount of Organic Matter in Soils   33

6 Let's Get Physical: Soil Tilth, Aeration, and Water   41

7 Nutrient Cycles and Flows   55

PART TWO

ECOLOGICAL SOIL & CROP MANAGEMENT

8   Managing for High Quality Soils   63

9   Animal Manures   77

10 Cover Crops   87

11 Crop Rotations   99

12 Making and Using Composts   109

13 Reducing Soil Erosion   119

14 Preventing and Lessening Compaction   125

15 Reducing Tillage   135

16 Nutrient Management: An Introduction   147

17 Management of Nitrogen and Phosphorus   157

18 Other Fertility Issues: Nutrients, CEC, Acidity and Alkalinity   167

19 Getting the Most from Soil Tests   177

PART THREE

PUTTING IT ALL TOGETHER

20 How Good are Your Soils? On-Farm Soil Health Evaluation   203

21 Putting it All Together   209

Glossary   215

Resources   221

Index   225

Preface

To use the land without abusing it.

— J. OTIS HUMPHRY, EARLY 1900s

We have written this book with farmers, extension agents, students, and gardeners in mind. *Building Soils for Better Crops* is a practical guide to ecological soil management that provides background information as well as details of soil-improving practices. This book is meant to give the reader an appreciation of the importance of soil health and to suggest ecologically sound practices that help to develop and maintain healthy soils.

The first edition of *Building Soils for Better Crops* focused exclusively on soil organic matter management. If you follow practices that build and maintain good levels of soil organic matter, you will find it easier to grow healthy and high-yielding crops. Plants can withstand droughty conditions better and won't be as bothered by insects and diseases. By maintaining adequate levels of organic matter in soil, there is less reason to use as much commercial fertilizer and lime as many farmers now purchase. Soil organic matter is that important!

Although organic matter management is the heart of the second edition, we decided to write a more comprehensive guide that includes the other essential aspects of building healthy soils. This edition contains four chapters, two new and two completely rewritten, on managing soil physical properties. We also included four new chapters on nutrient management and one on evaluating soil health. In addition, farmer profiles describe a number of key practices that enhance the health of their soils.

A book like this one cannot give exact answers to problems on specific farms. There are just too many differences from one field to another, and one farm to another, to warrant blanket recommendations. To make specific suggestions, it is necessary to know the details of the soil, crop, climate, machinery, human considerations, and other variable factors. Good soil management is better achieved through education and understanding than with blanket recommendations.

Over many centuries, people have struggled with the same issues we struggle with today. We quote some of these persons in epigraphs at the beginning of each chapter in appreciation for those who have come before. Vermont Agricultural Experiment Station Bulletin No. 35, published in 1908, is especially fascinating. It contains an article by three scientists about the importance of soil organic matter that is strikingly modern in many ways. Another example from more than a half century ago: The message of Edward Faulkner's *Plowman's Folly*, that reduced tillage and increased use of organic residues are essential to improving soil, is as valid today as in 1943 when it was first published. The saying is right — what goes around comes around. Sources cited at the end of chapters are those we referred to during writing. They are not a comprehensive list of references on the subject.

Many people reviewed individual chapters or the entire manuscript at one stage or another and made very useful suggestions. We would like to thank: Jim Bauder, Douglas Beegle, Keith Cassel, Andy Clark, Steve Diver, John Doran, Tim Griffin, Vern Grubinger, Wendy Sue Harper, John Hall, John Hart, Bill Jokela, Keith Kelling, Fred Kirschenmann, Shane LaBrake, Bill Liebhardt, Birl Lowery, Charles Mitchell, Paul Mugge, Cass Peterson, George Rehm, Joel Rissman, Eric Sideman, Ev Thomas, Michelle Wander, and Ray Weil. Special thanks to Valerie Berton, SARE communications specialist, who wrote the farm profiles, copyedited the manuscript and oversaw production. Any mistakes are, of course, ours alone.

— Fred Magdoff
Department of Plant & Soil Science
University of Vermont

& Harold van Es
Department of Crop & Soil Sciences
Cornell University
May 2000

Introduction

Used to be anybody could farm. All you needed was
a strong back. . . but nowadays you need a good
education to understand all the advice you get so
you can pick out what'll do you the least harm.

—VERMONT SAYING, MID-1900s

One of our truly modern miracles is our agricultural system, which produces abundant, affordable food. High yields come from the use of improved crop varieties, fertilizers, pest control products, and irrigation. At the same time, mechanization and the ever-increasing capacity of field equipment allows farmers to work increasing acreage.

Despite the high productivity per acre and per person, many farmers, agricultural scientists, and extension specialists see severe problems associated with our intensive agricultural production systems. Examples abound:

- Too much nitrogen fertilizer or animal manure sometimes causes high nitrate concentrations in groundwater. These concentrations can become high enough to pose a human health hazard.

- Phosphate in runoff water enters water bodies and degrades their waters by stimulating algae growth.

- Antibiotics used to fight diseases in farm animals can enter the food chain and may be found in the meat we eat. Their overuse has resulted in outbreaks of human illness from strains of disease-causing bacteria that have become resistant to antibiotics.

- Erosion associated with conventional tillage and lack of good rotations degrades our precious soil and, at the same time, causes the silting up of reservoirs, ponds, and lakes.

The food we eat and our surface and ground waters are sometimes contaminated with disease-causing organisms and chemicals used in agriculture. Pesticides used to control insects and plant diseases can be found in foods, animal feeds, groundwater, and in surface water running off agricultural fields. Farmers and farm workers are at special risk. Studies have shown higher cancer rates among those who work with or near certain pesticides. The general public is increasingly demanding safe, high quality food that is produced without excessive damage to the environment — and many are willing to pay a premium to obtain it.

Farmers are also in a perpetual struggle to maintain a decent standard of living. As consolidations and other changes occur in the agriculture input, food processing, and marketing sectors, the farmer's bargaining position weakens. The high cost of purchased inputs and the low prices of many agricultural commodities, such as wheat, corn, cotton, and milk, have caught farmers in a cost-price squeeze that makes it hard to run a profitable farm.

Given these problems, you might wonder if we should continue to farm in the same way. A major effort is underway to develop and implement practices that are both more environmentally sound than conventional practices and at the same time more economically rewarding for farmers. As farmers use management skills and better knowledge to work more closely with the biological world, they frequently find that there are ways to decrease use of products purchased off the farm.

With the new emphasis on sustainable agriculture comes a reawakening of interest in soil health. Early scientists, farmers, and gardeners were well aware of the importance of soil quality and organic matter to the productivity of soil. The significance of soil organic matter, including living organisms in the soil, was understood by scientists at least as far back as the 17th century. John Evelyn, writing in England during the 1670s, described the importance of topsoil and explained that the productivity of soils tended to be lost with time. He noted that their fertility could be maintained by adding organic residues. Charles Darwin, the great natural scientist of the 19th century who developed the modern theory of evolution, studied and wrote about the importance of earthworms to the cycling of nutrients and the general fertility of the soil.

. . . organic matter was
"once extolled as the essential
soil ingredient, the bright particular
star in the firmament of the
plant grower . . ."

Around the turn of the 20th century, there was again an appreciation of the importance of soil health. Scientists had realized that "worn out" soils, where productivity had drastically declined, resulted mainly from the depletion of soil organic matter. At the same time, they could see a transformation coming: although organic matter was "once extolled as the essential soil ingredient, the bright particular star in the firmament of the plant grower, it fell like Lucifer" under the weight of "modern" agricultural ideas (Hills, Jones, and Cutler, 1908). With the availability of inexpensive fertilizers and larger farm equipment after World War II, and of cheap water for irrigation in some parts of the western United States, many people working with soils forgot or ignored the importance of organic matter in promoting high quality soils.

As farmers and scientists placed less emphasis on soil organic matter during the last half of the 20th century, farm machinery was getting larger. More horse power for tractors allowed more land to be worked by fewer people. Large 4-wheel drive tractors allowed farmers to do field work when the soil was wet, creating severe compaction and sometimes leaving the soil in a cloddy condition, requiring more harrowing than otherwise would be needed. The use of the moldboard plow, followed by harrowing, broke down soil structure and left no residues on the surface. Soils were left bare and very susceptible to wind and water erosion. New harvesting machinery was developed, replacing hand-harvesting of crops. As dairy herd size increased, farmers needed bigger spreaders to handle the manure. The use of larger equipment created new problems. Making many passes through the field with heavy equipment for spreading fertilizer and manure, preparing a seedbed, planting, spraying pesticides, and harvesting created the potential for significant amounts of soil compaction.

What many people think are individual problems...may just be symptoms of a degraded, poor quality soil.

A new logic developed that most soil-related problems could be dealt with by increasing external inputs. This is a *reactive* way of dealing with soil issues — you react after seeing a "problem" in the field. If a soil is deficient in some nutrient, you buy a fertilizer and spread it on the soil. If a soil doesn't store enough rainfall, all you need is irrigation. If a soil becomes too compacted and water or roots can't easily penetrate, you use an implement, such as a subsoiler, to tear it open. If a plant disease or insect infestation occurs, you apply a pesticide.

Are low nutrient status, poor water-holding capacity, soil compaction, susceptibility to erosion, and disease, nematode, or insect damage really individual and unrelated problems? Perhaps they are better viewed as only symptoms of a deeper, underlying problem. The ability to tell the difference between what is the underlying problem and what is only a symptom of a problem is essential to deciding on the best course of action. For example, if you are hitting your head against a wall and you get a headache — is the problem the headache, and aspirin the best remedy? Clearly, the real problem is your behavior and not the headache, and the best solution is to stop banging your head on the wall!

What many people think are individual problems may just be symptoms of a degraded, poor quality soil. These symptoms are usually directly related to depletion of soil organic matter, lack of a thriving and diverse population of soil organisms, and compaction caused by use of heavy field equipment. Farmers have been encouraged to react to individual symptoms instead of focusing their attention on general soil health management. A new approach is needed to help develop farming practices that take advantage of the inherent strengths of natural systems. In this way, we can prevent the many symptoms of unhealthy soils from developing instead of reacting after they develop. If we are to work together with nature, instead of attempting to overwhelm and dominate it, the buildup and maintenance of good levels of organic matter in our soils is as critical as management of physical conditions, pH and nutrient levels.

This book has three parts. Part One provides background information about soil health and organic matter: what it is, why it is so important to general soil health and why some soils are of higher quality than others. Also included are discussions of soil physical properties, soil water storage, and nutrient cycles and flows. Part Two deals with practices that promote building better soils — with a lot of emphasis on promoting organic matter buildup and maintenance. Following practices that build and maintain organic matter may be the key to soil fertility and may help solve many problems. However, other soil-management practices also are needed to supplement soil organic matter management. Practices for enhancing soil quality include the use of animal manures and cover crops; good residue management; appropriate selection of rotation crops; use of composts; reduced tillage; minimizing soil compaction and enhancing aeration; better nutrient and amendment management; and adapting specific conservation practices for erosion control. Part Three deals with how to combine soil-building management strategies that actually work on the farm and how to evaluate soil to tell if its health is improving.

SOURCE
Hills, J.L., C.H. Jones, and C. Cutler. 1908. Soil deterioration and soil humus. pp. 142–177. In *Vermont Agricultural Experiment Station Bulletin 135*. College of Agriculture, University of Vermont. Burlington, Vermont.

PART ONE

The Basics of
Soil Organic Matter,
Physical Properties,
and Nutrients

1

Healthy Soils

All over the country [some soils are]
worn out, depleted, exhausted, almost dead.
But here is comfort: These soils possess possibilities
and may be restored to high productive power,
provided you do a few simple things.

—C.W. BURKETT, 1907

It should come as no surprise that many cultures have considered soil central to their lives. After all, people were aware that the food they ate grew from the soil. Our ancestors who first practiced agriculture must have been amazed to see life reborn each year when seeds placed in the ground germinated and then grew to maturity. In the Hebrew bible, the name given to the first man (Adam) is the masculine version of the word earth or soil (adama). The name for the first woman (Eve, or Hava in Hebrew) comes from the word for living. Soil and human life were considered to be intertwined. A particular reverence for the soil has been an important part of the cultures of many other civilizations, including American Indian tribes.

Although we focus on the critical role soils play in growing crops, it's important to keep in mind that soils also serve other important purposes. Soils govern what percent of the rainfall runs off the field, as compared to the percent that enters the soil and eventually helps recharge underground aquifers. When a soil is denuded of vegetation and it starts to degrade, excessive runoff and flooding are more common. Soils also absorb, release, and transform many different chemical compounds. For example, they help to purify wastes flowing from the septic system drain in your back yard. Soils also provide habitats for a diverse group of organisms, some of which are very important — such as with those bacteria that produce antibiotics. Soil organic matter stores a huge amount of atmospheric carbon. Carbon, in the form of carbon dioxide, is a greenhouse gas associated with global warming. We also build roads and buildings on soils; some are definitely better than others for this purpose.

3

WHAT KIND OF SOIL DO YOU WANT?

Farmers sometimes use the term *soil health* to describe the condition of the soil. Scientists usually use the term *soil quality*, but both refer to the same idea — how good is the soil in its role of supporting the growth of high yielding, healthy crops?

How would you know a high quality soil from a lower quality soil? Most farmers or gardeners would say that they know one when they see one. Farmers can certainly tell you which of the soils on their farms are of low, medium, or high quality. They know high quality soil because it generates higher yields with less effort. Less rainwater runs off and fewer visible signs of erosion are seen on the better quality soils. Less power is needed to operate machinery on a healthy soil than on poorer, compacted soils. Soil scientists are working together with farmers and agricultural extension personnel to try to come up with a widely accepted definition of soil health and to determine what factors (pH, bulk density, aggregate stability, etc.) need to be measured to estimate a soil's quality.

The first thing many might think of is that the soil should have a sufficient supply of nutrients throughout the growing season. But don't forget, at the end of the season there shouldn't be too much nitrogen and phosphorus left in highly soluble forms or enriching the soil's surface. Leaching and runoff of nutrients are most likely to occur after crops are harvested and before the following year's crops are well established.

We also want the soil to have good tilth so that plant roots can fully develop with the least amount of effort. A soil with good tilth is more spongy and less compact than a soil with poor tilth. A soil that has a favorable and stable soil structure also promotes rainfall infiltration and water storage for plants to use later. For good root growth and drainage, we also want a soil with sufficient depth before there's a restricting layer.

We want a soil to be well drained, so it dries enough to permit timely field operations. Also, it's essential that oxygen is able to reach the root zone to promote optimal root health — and that happens best in a soil without a drainage problem. (Keep in mind that these general characteristics do not hold for all crops. For example, flooded soils are important for paddy rice production.)

For soil thou art...

—Book of Genesis

We want the soil to have low populations of plant disease and parasitic organisms so plants grow better. Certainly, there should also be a low weed pressure, especially of aggressive and hard-to-control weeds. Most soil organisms are beneficial and we certainly want high amounts of organisms that help plant growth, such as earthworms and many bacteria and fungi.

A high quality soil is free of chemicals that might harm the plant. These can occur naturally, such as soluble aluminum in very acid soils or excess salts in arid region soils. Potentially harmful chemicals also are introduced by human activity, such as fuel oil spills or application of sewage sludge with high concentrations of toxic elements.

A high quality soil should resist being degraded. It also should be resilient, recovering quickly after unfavorable changes like compaction.

THE NATURE AND NURTURE OF SOILS

Some soils are exceptionally good for growing crops and others are inherently unsuitable; most are in between. Many soils also have limitations, such as low organic matter content, texture extremes (coarse sand or heavy clay), poor drainage, and layers that restrict root growth. Iowa's loess-derived prairie soils are naturally blessed with a combination of silt loam texture and high organic matter contents. By every standard for assessing soil health, these soils — in their virgin state — would rate very high. We can compare them with a person who is naturally very healthy and has great athletic abilities. Many of us are not quite so lucky and Nature has given us qualities that may never make us great baseball players, swimmers, or marathon runners, even if we tried very hard.

Good soil organic matter management is … the very foundation for a more sustainable and thriving agriculture.

The way we care for, or *nurture*, a soil modifies its inherent nature. A good soil can be abused through years of poor management and turn into one with poor health, although it generally takes a lot of mistreatment to reach that point. On the other hand, an innately challenging soil may be very "unforgiving" of poor management and quickly become even worse. For example, a heavy clay loam soil can be easily compacted and turn into a dense mass. Both the naturally good and poor soils can be productive if they are managed well. However, they will probably never reach parity, because some limitations simply cannot be completely overcome. The key idea, however, is the same that we wish for our children — we want our soils to reach their fullest potential.

HOW DO YOU BUILD A HEALTHY, HIGH QUALITY SOIL?

Some characteristics of healthy soils are relatively easy to achieve — for example, an application of limestone will make a soil less acid and increase availability of many nutrients to plants. But what if the soil is only a few inches deep? There is little that can be done within economic reason, except on a very small garden-size plot. If the soil is poorly drained because of a restricting subsoil layer of clay, tile drainage can be installed, but at a significant cost.

We use the term *building soils* to emphasize that the nurturing process of converting a degraded or low quality soil into a truly high quality one requires understanding, thought, and significant actions. This is also true for maintaining or improving already healthy soils. Soil organic matter influences almost all of the characteristics we've just discussed. For soil tilth, organic matter is one of the main influences. Organic matter is even critical for managing pests — and good management of this resource should be the starting point for a pest management program on every farm. Good organic matter management is, therefore, the foundation for high quality, healthy soils. Practices that promote good soil organic matter management are, thus, the very foundation for a more sustainable and thriving agriculture. It is for this reason that so much space is devoted to organic matter in this book. However, we cannot forget other critical aspects of management — such as trying to lessen compaction by heavy field equipment and good nutrient management.

Although the details of how best to create high quality soils differ from farm to farm and even field to field, the general approaches are the same:

- Use a number of practices that add organic materials to the soil.
- Use diverse sources of organic materials.
- Reduce unneeded losses of native soil organic matter.
- Use practices that leave the soil surface protected from raindrops and temperature extremes.
- Whenever traveling on the soil with field equipment, use practices that help develop and maintain good soil structure.
- Manage soil fertility status to maintain optimal pH levels for your crops and a sufficient

supply of nutrients for plants without resulting in water pollution.

- In arid regions, a combination of gypsum and leaching may be needed to reduce the amount of sodium or salt in the soil.

HOW DO SOILS BECOME DEGRADED?

Although we want to emphasize healthy, high quality soils, it is also crucial to recognize that many soils in the U.S. and around the world have become degraded — what many used to call "worn out" soils. Degradation most commonly occurs when erosion and decreased soil organic matter levels initiate a downward spiral (figure 1.1). Soils become compact, making it hard for water to infiltrate and roots to develop properly. Erosion continues and nutrients decline to levels too low for good crop growth. The development of saline (too salty) soils under irrigation in arid regions is another cause of

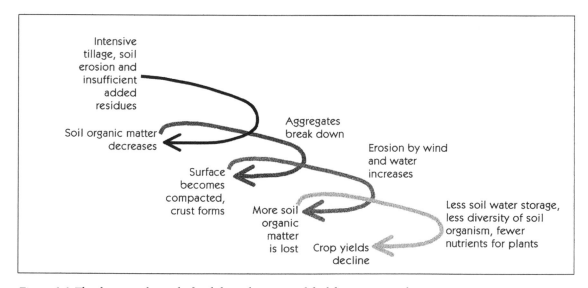

Figure 1.1 The downward spiral of soil degradation. Modified from Topp et al., 1995.

reduced soil health. (Salts added in the irrigation water need to be leached beneath the root zone to avoid the problem.)

Historically, soil degradation has caused significant harm to many early civilizations, including the drastic loss of productivity resulting from soil erosion in Greece and many locations in the Middle East (such as Israel, Jordan, and Lebanon). This led to either colonial ventures to help feed the citizenry or to the decline of the early cultures.

Tropical rainforest conditions (high temperature and rainfall, with most of the organic matter near the soil surface) may cause significant soil degradation within two or three years of conversion to cropland. This is the reason that the "slash and burn" system, with people moving to a new patch of forest every few years, developed in the tropics. After farmers depleted the soils in a field, they would cut down and burn the trees in the new patch, allowing the forest and soil to regenerate in previously cropped areas.

The westward push of U.S. agriculture was stimulated by rapid soil degradation in the East, originally a zone of temperate forest. Under the conditions of the humid portion of the Great Plains (moderate rainfall and temperature, with organic matter distributed deeper in the soil), it took many decades for the effects of soil degradation to become evident.

SOURCES

Doran, J.W., M. Sarrantonio, and M.A. Liebig. 1996. Soil heath and sustainability. pp. 1–54 In *Advances in Agronomy* Vol. 56. Academic Press, Inc. San Diego, CA.

Hillel, D. 1991. *Out of the Earth: Civilization and the Life of the Soil.* University of California Press. Berkeley, CA.

Spillman, W.J. 1906. *Renovation of Worn-out Soils. Farmers' Bulletin* No. 245. USDA, Government Printing Office. Washington, D.C.

Topp, G.C., K.C. Wires, D.A. Angers, M.R. Carter, J.L.B. Culley, D.A. Holmstrom, B.D. Kay, G.P. Lafond, D.R. Langille, R.A. McBride, G.T. Patterson, E. Perfect, V. Rasiah, A.V. Rodd, K.T. Webb. 1995. Changes in soil structure. Chapter 6. In *The Health of Our Soils: Toward Sustainable Agriculture in Canada* (Acton, D.F. and L.J. Gregorich (eds.). Centre for Land and Biological Resources Research. Research Branch, Agriculture and Agri-Food Canada. Publication 1906/E. http://res.agr.ca/CANSIS/PUBLICATIONS/HEALTH/_overview.html

2

What is Soil Organic Matter?

Soil consists of four important parts: mineral solids, water, air, and organic matter. Mineral solids are sand, silt, and clay. Sand has the largest particle size; clay has the smallest. The minerals mainly consist of silicon, oxygen, aluminum, potassium, calcium, and magnesium. The soil water, also called the soil solution, contains dissolved nutrients and is the main source of water for plants. Essential nutrients are made available to the roots of plants through the soil solution. The air in the soil, which is in contact with the air above ground, provides roots with oxygen and helps remove excess carbon dioxide from respiring root cells. The clumping together of mineral and organic particles to form aggregates of various sizes is a very important property of soils. Compared to poorly aggregated soils, those with good aggregation usually have better tilth and contain more spaces, or pores, for storing water and allowing gas exchange.

Organic matter has an overwhelming effect on almost all soil properties, although it is generally present in relatively small amounts. A typical agricultural soil has 1 to 6 percent organic matter. It consists of three distinctly different parts — living organisms, fresh residues, and well-decomposed residues. These three parts of soil organic matter have been described as the living, the dead, and the very dead. This three-way classification may seem simple and unscientific, but it is very useful.

The living part of soil organic matter includes a wide variety of microorganisms, such as bacteria, viruses, fungi, protozoa, and algae. It even includes plant roots and the insects, earthworms, and larger animals, such as moles, woodchucks, and rabbits, that spend some of their time in

Soil Organic Matter—
the living
the dead
the very dead

the soil. The living portion represents about 15 percent of the total soil organic matter. Microorganisms, earthworms, and insects help break down crop residues and manures and, as they use the energy of these materials, mix them with the minerals in the soil. In the process, they recycle plant nutrients. Sticky substances on the skin of earthworms and those produced by fungi help bind particles together. This helps to stabilize the soil *aggregates,* clumps of particles that make up good soil structure. Organisms such as earthworms and some fungi also help to stabilize the soil's structure (for example, by producing channels that allow water to infiltrate) and, thereby, improve soil water status and aeration. A good soil structure increases water filtering into the soil and decreases erosion. Plant roots also interact in significant ways with the various microorganisms and animals living in

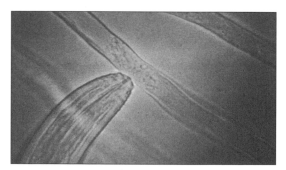

Figure 2.1 A nematode feeds on a fungus, part of a living system of checks and balances. Photo by Harold Jensen.

the soil. Another important aspect of soil organisms is that they are in a constant struggle with each other (figure 2.1). Further discussion of the interactions between soil organisms and roots, and among the various soil organisms, is provided in chapter 3.

The fresh residues, or "dead" organic matter, consist of recently deceased microorganisms, insects, earthworms, old plant roots, crop residues, and recently added manures. In some cases, just looking at them is enough to identify the origin of the fresh residues (figure 2.2). This part of soil organic matter is the active, or easily decomposed, fraction. This active fraction of soil organic matter is the main supply of food for various organisms living in the soil — microorganisms, insects, and earthworms. As organic materials decompose, they release many of the nutrients needed by plants. Organic chemical compounds produced during the decomposition of fresh residues also help to bind soil particles together and give the soil a good structure.

Organic molecules directly released from cells of fresh residues, such as proteins, amino acids, sugars, and starches, are also considered part of this fresh organic matter. These molecules generally do not last long in the soil because so many microorganisms use them as food.

The well-decomposed organic material in soil, the "very dead," is called *humus.* Humus is a term sometimes used to describe all soil organic matter. Some use it to describe just the part you can't see without a microscope. We'll use the term to refer only to the well-decomposed part of soil organic matter. The already well-decomposed humus is not a food for organisms, but its very small size and chemical properties make it an important part of the soil. Humus holds on to some essential nutrients, storing them for slow release to plants. Humus

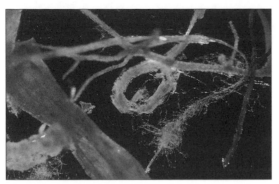

Figure 2.2 Partially decomposed fresh residues (the "dead") removed from soil. Fragments of stems, roots, fungal hyphae, are all readily used by soil organisms.

also can surround certain potentially harmful chemicals and prevent them from causing damage to plants. Good amounts of soil humus can both lessen drainage or compaction problems that occur in clay soils and improve water retention in sandy soils.

Organic matter decomposition is a process that is similar to the burning of wood in a stove. When burning wood reaches a certain temperature, the carbon in the wood combines with oxygen from the air and forms carbon dioxide. As this occurs, the energy stored in the carbon-containing chemicals in the wood is released as heat in a process called oxidation. The biological world, including humans, animals, and microorganisms, also makes use of energy inside carbon-containing molecules. This process of converting sugars, starches, and other compounds into a directly usable form of energy is also a type of oxidation. We usually call it *respiration*. Oxygen is used and carbon dioxide and heat are given off in this process.

A multitude of microorganisms, earthworms, and insects get their energy and nutrients by breaking down organic residues in soils. At the same time, much of the energy stored in residues is used by organisms to make new chemicals as well as new cells. How does energy get stored inside organic residues in the first place? Green plants use the energy of sunlight to link carbon atoms together into larger molecules. This process, known as *photosynthesis*, is used by plants to store energy for respiration and growth.

Soil carbon is sometimes used as a synonym for organic matter. Because carbon is the main building block of all organic molecules, the amount in a soil is very strongly related to the

---

Soil organic carbon is another way of referring to organic matter.

---

total amount of all the organic matter — the living organisms plus fresh residues plus well decomposed residues. However, under semiarid conditions, it is common to also have another form of carbon in soils — limestone either as round concretions or dispersed evenly throughout the soil. Lime is calcium carbonate, which contains calcium, carbon, and oxygen. This is an *inorganic* carbon form. Even in humid climates, when limestone is found very close to the surface, some may be present in the soil. So, when people talk about soil carbon instead of organic matter, they are usually referring to *organic* carbon. The amount of organic matter in soils is about twice the organic carbon level.

Source

Brady, N.C., and R.R. Weil. 1999. *The Nature and Properties of Soils*. 12th ed. Macmillan Publishing Co. New York, NY.

3

The Living Soil

*The plow is one of the most ancient and most valuable
of man's inventions; but long before he existed the
land was in fact regularly ploughed, and continues to
be thus ploughed by earthworms.*

—Charles Darwin, 1881

When soil organisms and roots go about their normal functions of getting energy for growth from organic molecules they "respire" — using oxygen and releasing carbon dioxide to the atmosphere. (Of course, as we take our essential breaths of air, we do the same.) An entire field can be viewed as breathing as if it is one large organism. The soil is like an organism in another way too — a field also may get "sick" in the sense that it becomes incapable of supporting healthy plants.

The organisms living in the soil, both large and small, play a significant role in maintaining a healthy soil system and healthy plants. One of the main reasons we are interested in these organisms is because of their role in breaking down organic residues and incorporating them into the soil. Soil organisms influence every aspect of decomposition and nutrient availability. As organic materials are decomposed, nutrients become available to plants, humus is produced, soil aggregates are formed, channels are created for water infiltration and better aeration, and those residues originally on the surface are brought deeper into the soil.

We classify soil organisms in several different ways. Each organism can be discussed separately or all organisms that do the same types of things can be discussed as a group. We also can look at soil organisms according to their role in the decomposition of organic materials. For example, organisms that use fresh residues as their source of food are called primary (1°), or first-level, consumers of organic materials (see figure 3.1). Many of these primary consumers break down large pieces of residues into smaller fragments. Secondary (2°) consumers are organisms that feed on the primary consumers themselves or their waste products. Tertiary (3°) consumers then feed on the secondary consumers. Another way to treat organisms is by general size, such as very small, small, medium, large, and very large. This is how we will discuss soil organisms in this chapter.

There is constant interaction among the organisms living in the soil. Some organisms help other organisms, as when bacteria that live inside the earthworm's digestive system help decompose organic matter. Although there are many examples of such mutually beneficial *symbiotic relationships*, an intense competition occurs among most of the diverse organisms in healthy soils. Organisms may directly compete with each other for the same food. Some organisms naturally feed on others — nematodes may feed on fungi, bacteria, or other nematodes, and some fungi trap and kill nematodes.

Some soil organisms can harm plants either by causing disease or by being parasites. In other words, there are "good" as well as "bad" bacteria, fungi, nematodes, and insects. One of the goals of agricultural production systems should be to create conditions that enhance the growth of beneficial organisms, which are the vast majority, while decreasing populations of those few that are potentially harmful.

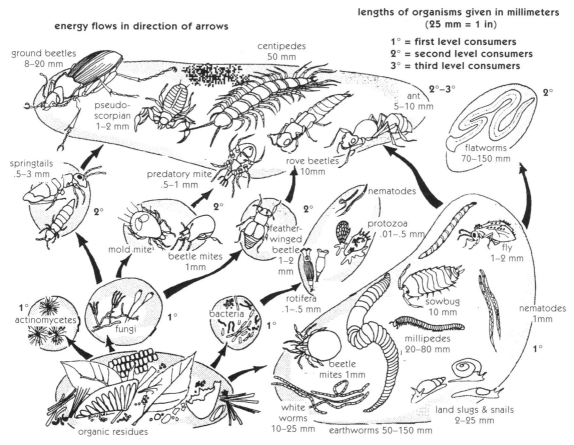

Figure 3.1 Soil organisms and their role in decomposing residues. Modified from D.L.Dindal, 1978.

SOIL MICROORGANISMS

Microorganisms are very small forms of life that can sometimes live as single cells, although many also form colonies of cells. A microscope is usually needed to see individual cells of these organisms. Many more microorganisms exist in topsoil, where food sources are plentiful, than in subsoil. They are especially abundant immediately next to plant roots, where sloughed off cells and chemicals released by roots provide ready food sources. These organisms are important primary decomposers of organic matter, but they do other things, such as providing nitrogen through fixation to help growing plants. Soil microorganisms have had another direct importance for humans — they are the origin of most of the antibiotic medicines we use to fight various diseases.

Bacteria

Bacteria live in almost any habitat. They are found inside the digestive system of animals, in the ocean and fresh water, in compost piles (even at temperatures over 130°F), and in soils. They are very plentiful in soils; a single teaspoon of topsoil may contain more than 50 million bacteria. Although some kinds of bacteria live in flooded soils without oxygen, most require well-aerated soils. In general, bacteria tend to do better in neutral soils than in acid soils.

In addition to being among the first organisms to begin decomposing residues in the soil, bacteria benefit plants by increasing nutrient availability. For example, many bacteria dissolve phosphorus, making it more available for plants to use.

Bacteria are also very helpful in providing nitrogen to plants. Although nitrogen is needed in large amounts by plants, it is often deficient in agricultural soils. You may wonder how soils can be deficient in nitrogen when we are surrounded by it — 78 percent of the air we breathe is composed of nitrogen gas. Yet plants as well as animals face the dilemma of the Ancient Mariner, who was adrift at sea without fresh water: "Water, water, everywhere nor any drop to drink." Unfortunately, neither animals nor plants can use nitrogen gas (N_2) for their nutrition. However, some types of bacteria are able to take nitrogen gas from the atmosphere and convert it into a form that plants can use to make amino acids and proteins. This conversion process is known as nitrogen fixation.

The microscopic flora and fauna in our soils give soil its fertility, otherwise it is just dirt.

—John Malcom, Vermont dairy farmer

Some nitrogen-fixing bacteria form mutually beneficial associations with plants. One such symbiotic relationship that is very important to agriculture is the nitrogen-fixing rhizobia group of bacteria that live inside nodules formed on the roots of legumes. These bacteria provide nitrogen in a form that leguminous plants can use, while the legume provides the bacteria with sugars for energy.

People eat some legumes or their products, such as peas, dry beans, and tofu made from soybeans. Soybeans, alfalfa, and clover are used for animal feed. Clovers and hairy vetch are grown as *cover crops* to enrich the soil with organic matter, as well as nitrogen, for the following crop. In an alfalfa field, the bacteria may fix hundreds of pounds of nitrogen per acre each year. With peas, the amount of nitrogen fixed is much lower, around 30 to 50 pounds per acre.

The *actinomycetes,* another group of bacteria, break large lignin molecules into smaller sizes. Lignin is a large and complex molecule found in plant tissue, especially stems, that is difficult for most organisms to break down. Lignin also frequently protects other molecules like cellulose from decomposition. Actinomycetes have some characteristics similar to fungi, but are sometimes grouped by themselves and given equal billing with bacteria and fungi.

Fungi

Fungi are another important type of soil microorganism. Yeast is a fungus used in baking and in the production of alcohol. A number of antibiotics are produced by other fungi. We have all probably let a loaf of bread sit around too long only to find fungus growing on it. We have seen or eaten mushrooms, the fruiting structure of some fungi. Farmers know that many plant diseases, such as downy mildew, damping-off, various types of root rot, and apple scab, are caused by fungi. Fungi are also important for starting the decomposition of fresh organic residues. They help get things going by softening organic debris and making it easier for other organisms to join in the decomposition process. Fungi are also the main decomposers of lignin. Fungi are less sensitive to acid-soil conditions than are bacteria. None are able to function without oxygen.

Many plants develop a beneficial relationship with fungi that increases the contact of roots with the soil. Fungi infect the roots and send out root-like structures called *hyphae* (see figure 3.2). The

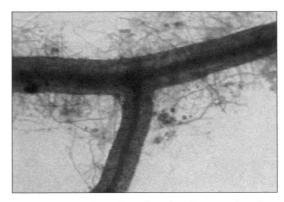

Figure 3.2 *Root heavily infected with mycorrhizal fungi (note round spores at the end of some hyphae). Photo by Sara Wright.*

hyphae of these *mycorrhizal* fungi take up water and nutrients that can then feed the plant. This is especially important for phosphorus nutrition of plants in low-phosphorus soils. The hyphae help the plant absorb water and nutrients and in return the fungi receive energy in the form of sugars, which the plant produces in its leaves and sends down to the roots. This symbiotic interdependency between fungi and roots is

Mycorrhizal Fungi

Mycorrhizal fungi help plants take up nutrients, improve nitrogen fixation by legumes, and help to form and stabilize soil aggregates. Crop rotations select for more types and better performing fungi than does mono-cropping. Some studies indicate that using cover crops, especially legumes, between main crops helps maintain high levels of spores and promotes good mycorrhizal development in the next crop. Roots that have lots of mycorrhizae are better able to resist fungal diseases, parasitic nematodes, and drought.

called a mycorrhizal relationship. All things considered, it's a pretty good deal for both the plant and the fungus. The hyphae of these fungi help develop and stabilize soil aggregates by secreting a sticky gel that glues mineral and organic particles together.

Fungi help to start the decomposition of organic residues.

Algae

Algae, like crop plants, convert sunlight into complex molecules like sugars, which they can use for energy and to help build the other molecules they need. Algae are found in abundance in the flooded soils of swamps and rice paddies. They also can be found on the surface of poorly drained soils or in wet depressions. Algae also occur in relatively dry soils, and they form mutually beneficial relationships with other organisms. Lichens found on rocks are an association between a fungus and an alga.

Protozoa

Protozoa are single-celled animals that use a variety of means to move about in the soil. Like bacteria and many fungi, they can be seen only with the help of a microscope. They are mainly secondary consumers of organic materials, feeding on bacteria, fungi, other protozoa, and organic molecules dissolved in the soil water. Protozoa — through their grazing on nitrogen-rich organisms and excreting wastes — are believed responsible for mineralizing much of the nitrogen (released from organic molecules) in agricultural soils.

SMALL AND MEDIUM-SIZED SOIL ANIMALS

Nematodes

Nematodes are simple soil animals that resemble tiny worms. They tend to live in the water films around soil aggregates. Some types of nematodes feed on plant roots and are well known plant pests. Diseases such as pythium and fusarium, which enter feeding wounds on the root, sometimes cause more damage than the feeding itself. However, most nematodes help in the breakdown of organic residues and feed on fungi, bacteria, and protozoa as secondary consumers. In fact, as with the protozoa, nematodes feeding on fungi and bacteria also helps convert nitrogen into forms for plants to use. As much as 50 percent or more of mineralized nitrogen comes from nematode feeding.

Earthworms

Earthworms are every bit as important as Charles Darwin believed more than a century ago. They are keepers and restorers of soil fertility. Different types of earthworms, including the night-crawler, field (garden) worm, and manure (red) worm, have different feeding habits. Some feed on plant residues that remain on the soil surface, while other types tend to feed on organic matter that is already mixed with the soil.

The surface-feeding nightcrawlers fragment and mix fresh residues with soil mineral particles, bacteria, and enzymes in their digestive system. The resulting material is given off as worm casts. Worm casts are generally higher in available plant nutrients, such as nitrogen, calcium, magnesium, and phosphorus, than the surrounding soil and, therefore, make an important contribution to the nutrient needs of plants. They also bring food down into their burrows, thereby mixing organic matter deep into the soil. Earthworms feeding on debris already below the surface continue to decompose organic materials and mix them with the soil minerals.

A number of types of earthworms, including the surface-feeding nightcrawler, make burrows that allow rainfall to easily infiltrate into the soil. These worms usually burrow to three feet or more under dry conditions. Even those types of worms that don't normally produce channels to the surface help loosen the soil, creating channels and cracks below the surface that help aeration and root growth. The number of earthworms in the soil ranges from close to zero to over a million per acre. Just imagine, if you create the proper conditions for earthworms, you can have 800,000 small channels per acre that conduct water into your soil during downpours.

Earthworms do some unbelievable work. They move a lot of soil from below up to the surface — from about 1 to 100 tons per acre each year. One acre of soil 6 inches deep weighs about 2 million pounds, or 1,000 tons. So 1 to 100 tons is the equivalent of about .006 of an inch to about half an inch of soil. Agricultural soils that use conservation practices may still erode, though at low annual rates of 1 to 4 tons/acre. A healthy earthworm population can counteract some of the effects of erosion — creating topsoil by bringing up subsoil and mixing it with organic residues.

Earthworms do best in well-aerated soils that are supplied with plentiful amounts of organic matter. A study in Georgia showed that soils with higher amounts of organic matter contained higher numbers of earthworms. Surface feeders, a type we would especially like to encour-

age, need residues left on the surface. They are harmed by plowing or disking, which disturbs their burrows and buries their food supplies. Worms are usually more plentiful under no-till practices than under conventional tillage systems. Although many pesticides have little effect on worms, others, such as aldicarb, parathion, and heptachlor, are very harmful to earthworms.

Diseases or insects that overwinter on leaves of crops can sometimes be partially controlled by high earthworm populations. The apple scab fungus — a major pest of apples in humid regions — and some leaf miner insects can be partly controlled when worms eat the leaves and incorporate the residues deeper into the soil.

Insects and Other Small to Medium-Sized Animals

Insects are another group of animals that inhabit soils. Common types of soil insects include termites, springtails, ants, fly larvae, and beetles. Many insects are secondary and tertiary consumers. Springtails feed on fungi and animal remains. Many beetles, in particular, eat other types of soil animals. Some beetles feed on weed seeds in the soil. Termites, well-known feeders of woody material, also consume decomposed organic residues in the soil.

Other medium- to large-sized soil animals include millipedes, centipedes, mites, slugs, snails, and spiders. Millipedes are primary consumers of plant residues, whereas centipedes tend to feed on other organisms. Mites may feed on food sources like fungi, other mites, and insect eggs, although some feed directly on residues. Spiders feed mainly on insects and their role in keeping insect pests from developing large populations can be important.

VERY LARGE ANIMALS

Very large animals, such as moles, rabbits, wood-chucks, snakes, prairie dogs, and badgers, burrow in the soil and spend at least some of their lives below ground. Moles are secondary consumers, with their diet consisting mainly of earthworms. Most of the other animals exist on vegetation. In many cases, their presence is considered a nuisance for agricultural production or lawns and gardens. Nevertheless, their burrows may help conduct water away from the surface during downpours and thus decrease erosion. In the South, the burrowing action of crawfish, abundant in many of the somewhat poorly drained soils, can have a large effect on soil structure. (In Texas and Louisiana, some rice fields are "rotated" with crawfish production.)

PLANT ROOTS

Healthy plant roots are essential for good crop yields. Roots are clearly influenced by the soil in which they live. If the soil is compact, low in nutrients or water, or has other problems, plants will not grow well. On the other hand, plants

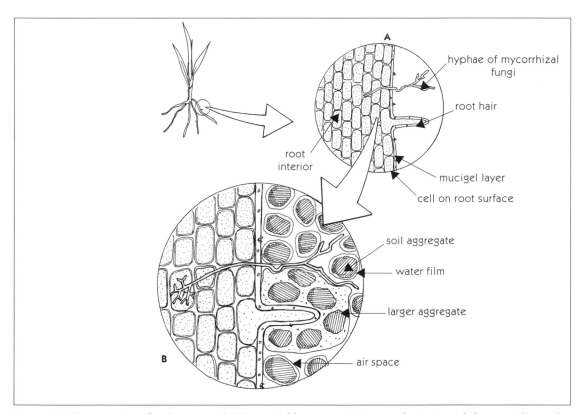

Figure 3.3 Close-up view of a plant root. A) The mucigel layer containing some bacteria and clay particles on the outside of the root. Also shown is a mycorrhizal fungus sending out its rootlike hyphae into the soil. B) Soil aggregates surrounded by thin films of water. Plant roots take water and nutrients from these films. Also shown is a larger aggregate made up of smaller aggregates pressed together and held in place by the root and hyphae.

also influence the soil in which they grow. The physical pressure of roots growing through soil helps form aggregates by bringing particles closer together. Small roots also help bind particles together. In addition, many organic compounds are given off, or exuded, by plant roots and provide nourishment for soil organisms living on or near the roots. A sticky layer surrounding roots, called the *mucigel,* provides very close contact between microorganisms, soil minerals, and the plant (figure 3.3).

For plants with extensive root systems, such as grasses, the amount of living tissue belowground may actually weigh more than the amount of leaves and stems we see above ground.

BIOLOGICAL DIVERSITY AND BALANCE

A diverse biological community in soils is essential to maintaining a healthy environment for plants. There may be over 100,000 different types of organisms living in soils. Of those, only a small number of bacteria, fungi, insects, and nematodes might harm plants in any given year. Diverse populations of soil organisms maintain a system of checks and balances that can keep disease organisms or parasites from becoming major plant problems. Some fungi kill nematodes and others kill insects. Still others produce antibiotics that kill bacteria. Protozoa feed on bacteria. Some bacteria kill harmful insects. Many protozoa, springtails, and mites feed on disease-causing fungi and bacteria. Beneficial organisms, such as the fungus *Trichoderma* and the bacteria *Pseudemonas fluorescens,* colonize plant roots and protect them from attack by harmful organisms. Some of these organisms, isolated from soils, are now sold commercially as biological control agents.

SOURCES

Alexander, M. 1977. *Introduction to Soil Microbiology.* 2d ed. John Wiley & Sons. New York, NY.

Hendrix, P.F., M.H. Beare, W.X. Cheng, D.C. Coleman, D.A. Crossley, Jr., and R.R. Bruce. 1990. Earthworm effects on soil organic matter dynamics in aggrading and degrading agroecosystems on the Georgia Piedmont. *Agronomy Abstracts,* p.250, American Society of Agronomy, Madison, WI.

Paul, E.A., and F.E. Clark. 1996. *Soil Microbiology and Biochemistry.* 2d ed Academic Press. San Diego, CA.

4

Why is Organic Matter So Important?

Why are soils which in our father's hands
were productive now relatively impoverished?

—J. L. Hills, C. H. Jones, and C. Cutler, 1908

Organic matter functions in a number of key roles to promote crop growth. It also is a critical part of a number of global and regional cycles.

A fertile soil is the basis for healthy plants, animals, and humans. Soil organic matter is the very foundation for healthy and productive soils. Understanding the role of organic matter in maintaining a healthy soil is essential for developing ecologically sound agricultural practices. It's true that you can grow plants on soils with little organic matter. In fact, you don't need any soil at all! [Although gravel or sand hydroponic systems without soil can grow excellent crops, large-scale systems of this type are usually neither economically or ecologically sound.] It's also true that there are other important issues aside from organic matter when considering the quality of a soil. However, as soil organic matter decreases, it becomes increasingly difficult to grow plants, because problems with fertility, water availability, compaction, erosion, parasites, diseases, and insects become more common. Ever higher levels of inputs — fertilizers, irrigation water, pesticides, and machinery — are required to maintain yields in the face of organic matter depletion. But if attention is paid to proper organic matter management, the soil can support a good crop without the need for expensive fixes.

The organic matter content of agricultural topsoil is usually in the range of 1 to 6 percent. A study of soils in Michigan demonstrated potential crop-yield increases of about 12 percent for every 1 percent organic matter. In a Maryland experiment, researchers saw an increase of approximately 80 bushels of corn per acre when organic matter increased from 0.8 to 2 percent.

What Makes Topsoil?

Having a good amount of topsoil is important. But what gives topsoil its beneficial characteristics? Is it because it's on TOP? If we bring in a bulldozer and scrape off one foot of soil, will the exposed subsoil now be topsoil because it's on the surface? Of course, everyone knows that there's more to topsoil than its location on the soil surface. Most of the properties we associate with topsoil — good nutrient supply, tilth, drainage, aeration, water storage, etc. — are there because topsoil is rich in organic matter and contains a huge diversity of life.

You might wonder how something that's only a small part of the soil can be so important for growing healthy and high-yielding crops. The enormous influence of organic matter on so many of the soil's properties — biological, chemical, and physical — makes it of critical importance to healthy soils (figure 4.1). Part of the explanation for this influence is the small particle size of the well-decomposed portion of organic matter — the humus. Its large surface area-to-volume ratio means that humus is in contact with a considerable portion of the soil. The intimate contact of humus with the rest of the soil allows many reactions, such as the release of available nutrients into the soil water,

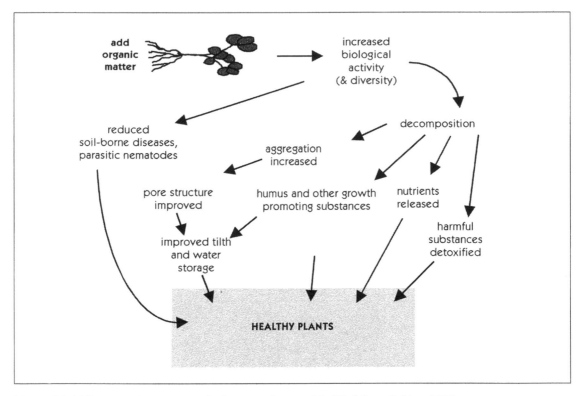

Figure 4.1 Adding organic matter results in many changes. Modified from Oshins, 1999.

BUILDING SOILS FOR BETTER CROPS

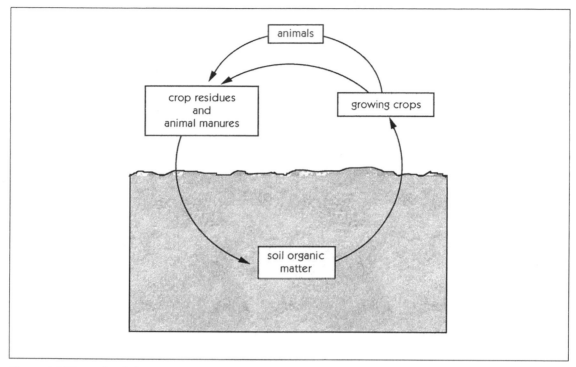

Figure 4.2 The cycle of plant nutrients.

to occur rapidly. However, the many roles of living organisms make soil life an essential part of the organic matter story.

Plant Nutrition

Plants need 18 chemical elements for their growth — carbon (C), hydrogen (H), oxygen (O), nitrogen (N), phosphorus (P), potassium (K), sulfur (S), calcium (Ca), magnesium (Mg), iron (Fe), manganese (Mn), boron (B), zinc (Zn), molybdenum (Mo), nickel (Ni), copper (Cu), cobalt (Co), and chlorine (Cl). Plants obtain carbon as carbon dioxide (CO_2) and oxygen partially as oxygen gas (O_2) from the air. The remaining essential elements are obtained mainly from the soil. The availability of these nutrients is influenced either directly or indirectly by the presence of organic matter. The elements needed in large amounts — carbon, hydrogen, oxygen, nitrogen, phosphorus, potassium, calcium, magnesium, sulfur — are called macronutrients. The other elements, called micronutrients, are essential elements needed in small amounts.

Nutrients from decomposing organic matter. Most of the nutrients in soil organic matter can't be used by plants as long as they exist as part of large organic molecules. As soil organisms decompose organic matter, nutrients are converted into simpler, inorganic, or mineral forms that plants can easily use. This process, called mineralization, provides much of the nitrogen that plants need by converting it from organic forms. For example, proteins are converted to ammonium (NH_4^+) and then to nitrate (NO_3^-). Most plants will take up the majority of their nitrogen from soils in the form of nitrate. The mineralization of organic matter is also an important mechanism for supplying plants with such nutrients as phosphorus and sulfur, and most of the micronutrients. This release of nutrients from organic matter by mineralization is part of a larger agricultural nutrient cycle (see figure 4.2). For a more detailed discussion of nutrient cycles and how they function in various cropping systems, see chapter 7.

Addition of nitrogen. Bacteria living in nodules on legume roots convert nitrogen from atmospheric gas (N_2) to forms that the plant can use directly. There are a number of free-living bacteria that also fix nitrogen.

Storage of nutrients on soil organic matter. Decomposing organic matter can feed plants directly, but it also can indirectly benefit the nutrition of the plant. A number of essential nutrients occur in soils as positively charged molecules called cations (pronounced cat-eye-ons). The ability of organic matter to hold onto cations in a way that keeps them available to plants is known as cation exchange capacity (CEC). Humus has many negative charges. Because opposite charges attract, humus is able to hold onto positively charged nutrients, such as calcium (Ca^{++}), potassium (K^+), and magnesium (Mg^{++}) (see figure 4.3a). This keeps them from leaching deep into the subsoil when water moves through the topsoil. Nutrients held in this way can be gradually released into the soil solution and made available to plants throughout the growing season. However, keep in mind that not all plant nutrients occur as cations. For example, the nitrate form of nitrogen is negatively charged (NO_3^-) and is actually repelled by the negatively charged CEC. Therefore, nitrate leaches easily as water moves down through the soil and beyond the root zone.

Clay particles also have negative charges on their surfaces (figure 4.3b), but organic matter may be the major source of negative charges for coarse and medium textured soils. Some types of clays, such as those found in the southeastern United States and in the tropics, tend to have low amounts of negative charge. When these clays are present, organic matter may be the major source of negative charges that bind nutrients, even for fine textured (high clay content) soils.

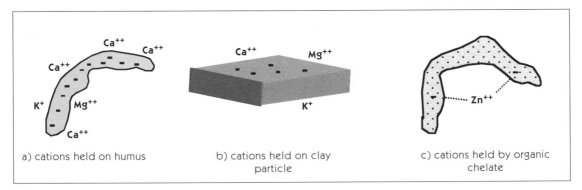

a) cations held on humus   b) cations held on clay particle   c) cations held by organic chelate

Figure 4.3 Cations held on organic matter and clay.

Protection of nutrients by chelation. Organic molecules in the soil may also hold onto and protect certain nutrients. These particles, called "chelates" (pronounced key-lates) are byproducts of active decomposition of organic materials and are smaller than those that make up humus. In general, elements are held more strongly by chelates than by binding of positive and negative charges. Chelates work well because they bind the nutrient at more than one location on the organic molecule (figure 4.3c). In some soils, trace elements, such as iron, zinc, and manganese, would be converted to unavailable forms if they were not bound by chelates. It is not uncommon to find low organic matter soils or exposed subsoils deficient in these micronutrients.

Other ways of maintaining available nutrients. There is some evidence that organic matter in the soil can inhibit the conversion of available phosphorus to forms that are unavailable to plants. One explanation is that organic matter coats the surfaces of minerals that can bond tightly to phosphorus. Once these surfaces are covered, available forms of phosphorus are less likely to react with them. In addition, humic substances may chelate aluminum and iron, both of which can react with phosphorus in the soil solution. When they are held as chelates, these metals are unable to form an insoluble mineral with phosphorus.

Beneficial Effects of Soil Organisms

Soil organisms are essential for keeping plants well supplied with nutrients because they break down organic matter. These organisms make nutrients available by freeing them from organic molecules. Some bacteria fix nitrogen gas from the atmosphere, making it available to plants. Other organisms dissolve minerals and make phosphorus more available. If soil organisms aren't present and active, more fertilizers will be needed to supply plant nutrients.

A varied community of organisms is your best protection against major pest outbreaks and soil fertility problems. A soil rich in organic matter and continually supplied with different types of fresh residues is home to a much more diverse group of organisms than soil depleted of organic matter. This greater diversity of organisms helps insure that fewer potentially harmful organisms will be able to develop sufficient populations to reduce crops yields.

Soil Tilth

When soil has a favorable physical condition for growing plants, it is said to have good tilth. Such a soil is porous and allows water to enter easily, instead of running off the surface. More water is stored in the soil for plants to use between rains and less soil erosion occurs. Good tilth also means that the soil is well aerated. Roots can easily obtain oxygen and get rid of carbon dioxide. A porous soil does not restrict root development and exploration. When a soil has poor tilth, the soil's structure deteriorates and soil aggregates break down, causing increased compaction and decreased aeration and water storage. A soil layer can become so compacted that roots can't grow. A soil with excellent physical properties will have numerous channels and pores of many different sizes.

Studies on both undisturbed and agricultural soils show that as organic matter increases, soils tend to be less compact and have more space for air passage and water storage. Sticky substances are produced during the decomposition of plant residues. Along with plant roots and fungal hyphae, they bind mineral particles together into clumps, or aggregates. In addition, the sticky secretions of mycorrhizal fungi — those that infect roots and help plants get more water and nutrients — are important binding material in soils. The arrangement and collection of minerals as aggregates and the degree of soil compaction have huge effects on plant growth (see chapter 6). The development of aggregates is desirable in all types of soils because it promotes better drainage, aeration, and water storage. The one exception is for wetland crops, such as rice, when you want a dense, puddled soil to keep it flooded.

Organic matter, as residue on the soil surface or as a binding agent for aggregates near the surface, plays an important role in decreasing soil erosion. Surface residues intercept raindrops and decrease their potential to detach soil particles. These surface residues also slow water as it flows across the field, giving it a better chance to infiltrate into the soil. Aggregates and large channels greatly enhance the ability of soil to conduct water from the surface into the subsoil.

Most farmers can tell that one soil is better than another by looking at them, touching them, how they work up when tilled, or even by sensing how they feel when walked on. What they are seeing or sensing is really good tilth. For an example, see the photo on the back cover of this book. It shows that soil differences can be created by different management strategies. Farmers and gardeners would certainly rather grow their crops on the more porous soil depicted in the photo on the right.

Since erosion tends to remove the most fertile part of the soil, it can cause a significant reduction in crop yields. In some soils, the loss of just a few inches of topsoil may result in a yield reduction of 50 percent. The surface of some soils low in organic matter may seal over, or crust, as rainfall breaks down aggregates, and pores near the surface fill with solids. When this happens, water that can't infiltrate into the soil runs off the field, carrying valuable topsoil (figure 4.4).

Large soil pores, or channels, are very important because of their ability to allow a lot of

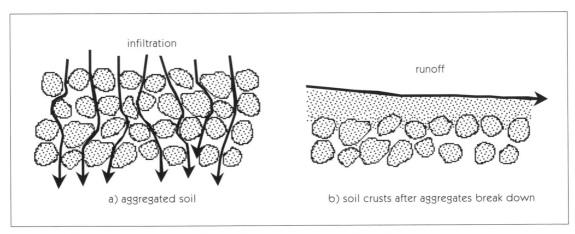

a) aggregated soil

b) soil crusts after aggregates break down

Figure 4.4 Changes in soil surface and water-flow pattern due to soil crusting.

water to flow rapidly into the soil. Larger pores are formed a number of ways. Old root channels may remain open for some time after the root decomposes. Larger soil organisms, such as insects and earthworms, create channels as they move through the soil. The mucus that earthworms secrete to keep their skin from drying out also helps to keep their channels open for a long time.

Figure 4.5 Corn grown in nutrient solution with (right) and without (left) humic acids. In this experiment by Rich Bartlett and Yong Lee, adding humic acids to a nutrient solution increased the growth of tomatoes and corn and increased the number and branching of roots. Photo by R. Bartlett.

Protection of the Soil Against Rapid Changes in Acidity

Acids and bases are released as minerals dissolve and organisms go about their normal functions of decomposing organic materials or fixing nitrogen. Acids or bases are excreted by the roots of plants, and acids form in the soil from the use of nitrogen fertilizers. It is best for plants if the soil acidity status, referred to as pH, does not swing too wildly during the season. The pH scale is a way of expressing the amount of free hydrogen (H^+) in the soil water. More acidic conditions, with greater amounts of hydrogen, are indicated by lower numbers. A soil at pH 4 is very acid. Its solution is 10 times more acid than a soil at pH 5. A soil at pH 7 is neutral — there is just as much base in the water as there is acid. Most crops do best when the soil is slightly acid and the pH is around 6 to 7. Essential nutrients are more available to plants in this pH range than when soils are either more acidic or more basic. Soil organic matter is able to slow down, or buffer, changes in pH by taking free hydrogen out of solution as acids are produced or by giving off hydrogen as bases are produced. (For discussion about management of acidic soils, see chapter 18.)

Stimulation of Root Development

Microorganisms in soils produce numerous substances that stimulate plant growth. Humus itself has a directly beneficial effect on plants (figure 4.5). Although the reasons for this stimulation are not yet understood, certain types of humus cause roots to grow longer and have more branches, resulting in larger and healthier plants. In addition, soil microorganisms produce a variety of root-stimulating substances that behave as plant hormones.

Darkening of the Soil

Organic matter tends to darken soils. You can easily see this in coarse-textured sandy soils containing light-colored minerals. Under well-drained conditions, a darker soil surface allows a soil to warm up a little faster in the spring. This provides a slight advantage for seed germination and the early stages of seedling development, which is often beneficial in cold regions.

Protection Against Harmful Chemicals

Some naturally occurring chemicals in soils can harm plants. For example, aluminum is an important part of many soil minerals and, as such, poses no threat to plants. As soils become more acidic, especially at pH levels below 5.5, aluminum becomes soluble. Some soluble forms of aluminum, if present in the soil solution, are toxic to plant roots. However, in the presence of significant quantities of soil organic matter, the aluminum is bound tightly and will not do as much damage.

Organic matter is the single most important soil property that reduces pesticide leaching. It holds tightly onto a number of pesticides. This prevents or reduces leaching of these chemicals into groundwater and allows time for detoxification by microbes. Microorganisms can change the chemical structure of some pesticides, industrial oils, many petroleum products (gas and oils), and other potentially toxic chemicals, rendering them harmless.

ORGANIC MATTER AND NATURAL CYCLES

The Carbon Cycle

Soil organic matter plays a significant part in a number of global cycles. People have become more interested in the carbon cycle because the buildup of carbon dioxide (CO_2) in the atmosphere is thought to cause global warming. Carbon dioxide is also released to the atmosphere when fuels, such as gas, oil, and wood, are burned. A simple version of the natural carbon cycle, showing the role of soil organic matter, is given in figure 4.6. Carbon dioxide is removed from the atmosphere by plants and used to make all the organic molecules necessary for life.

Each percent organic matter in the top 6 inches of soil contains about the same quantity of carbon as all the atmosphere directly over the field!

Sunlight provides plants with the energy they need to carry out this process. Plants, as well as the animals feeding on plants, release carbon dioxide back into the atmosphere as they use organic molecules for energy.

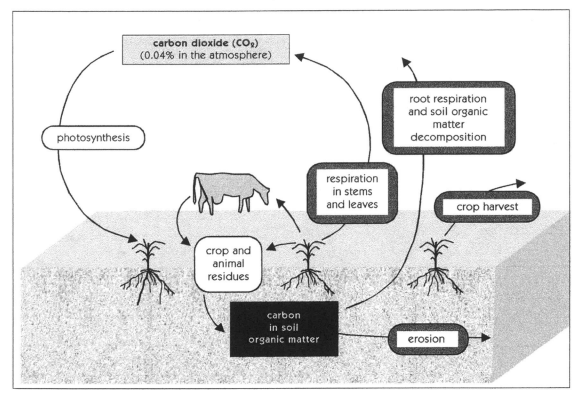

Figure 4.6 The role of soil organic matter in the carbon cycle. Losses of carbon from the field are indicated by the dark border around the words describing the process.

The largest amount of carbon present on the land is not in the living plants, but in soil organic matter. That is rarely mentioned in discussions of the carbon cycle. More carbon is stored in soils than in all plants, all animals and the atmosphere. Soil organic matter contains an estimated four times as much carbon as living plants. As soil organic matter is depleted, it becomes a source of carbon dioxide for the atmosphere. When forests are cleared and burned, a large amount of carbon dioxide is released to the atmosphere. There is a potentially larger release of carbon dioxide following conversion of forests to agricultural practices that rapidly deplete the soil of its organic matter. There is as much carbon in 6 inches of soil with 1 percent organic matter as there is in the atmosphere above a field. If organic matter decreases from 3 percent to 2 percent, the amount of carbon dioxide in the atmosphere could double. (Of course, wind and diffusion move the carbon dioxide to other parts of the globe.)

The Nitrogen Cycle

Another important global cycle in which organic matter plays a major role is the nitrogen cycle. This cycle is of direct importance in agriculture, because available nitrogen for plants is commonly deficient in soils. Figure 4.7 shows the nitrogen cycle and how soil organic matter enters into the cycle. Some bacteria living in soils are able to "fix" nitrogen, converting nitrogen gas to forms that other organisms, including crop plants, can use. Inorganic forms of nitrogen, like ammonium and nitrate, exist in the atmosphere naturally, although air pollution causes higher amounts than normal. Rainfall and snow deposit inorganic nitrogen forms on the soil. Inorganic nitrogen also may be added in the form of commercial nitrogen fertilizers. These fertilizers are derived from nitrogen gas in the atmosphere through an industrial fixation process.

Almost all of the nitrogen in soils exists as part of the organic matter, in forms that plants are not able to use as their main nitrogen source.

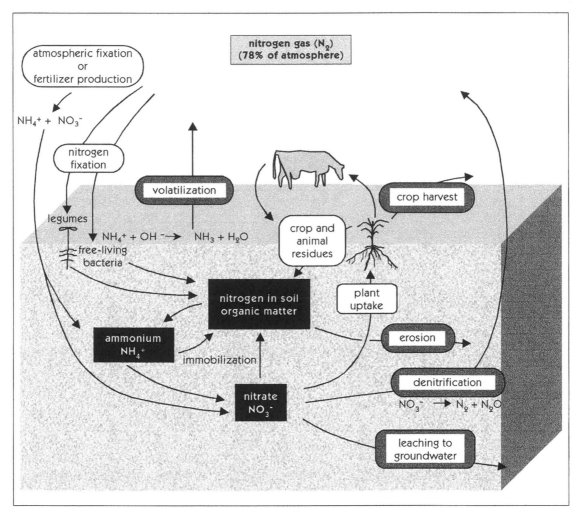

Figure 4.7 The role of soil organic matter in the nitrogen cycle. Losses of nitrogen from the field are indicated by the dark border around the words describing the process.

Bacteria and fungi convert the organic forms of nitrogen into ammonium and different bacteria convert ammonium into nitrate. Both nitrate and ammonium can be used by plants.

Almost all of the nitrogen in soils exists as part of the organic matter, in forms that plants are not able to use as their main nitrogen source.

Nitrogen can be lost from a soil in a number of ways. When crops are removed from fields, nitrogen and other nutrients also are removed. The nitrate (NO_3^-) form of nitrogen leaches readily from soils and may end up in groundwater at higher concentrations than may be safe for drinking. Organic forms of nitrate as well as nitrate and ammonium (NH_4^+) may be lost by runoff water and erosion. Once freed from soil organic matter, nitrogen may be converted to forms that end up back in the atmosphere. Bacteria convert nitrate to nitrogen (N_2) and nitrous oxide (N_2O) gases in a process called denitrification, which occurs in saturated soils. Nitrous oxide (it's called a "greenhouse gas") contributes to global warming. In addition, when it reaches the upper atmosphere, it helps to decrease the ozone levels that protect the earth's surface from the harmful effects of ultraviolet (UV) radiation. So if you needed another reason not to apply excessive rates of fertilizers or manures — in addition to the economic costs and the pollution of ground and surface waters — the possible formation of nitrous oxide should make you cautious.

The Water Cycle

Organic matter plays an important part in the local, regional, and global water, or hydrologic, cycle due to its role in promoting water infiltration into soils and storage within the soil. Water evaporates from the soil surface and from living plant leaves as well as from the ocean and lakes. Water then returns to the earth, usually far from where it evaporated, as rain and snow. Soils high in organic matter, with excellent tilth, enhance the rapid infiltration of rainwater into the soil. This water may be available for plants to use or it may percolate deep into the subsoil and help to recharge the groundwater supply. Since groundwater is commonly used as a drinking water source for homes and for irrigation, recharging groundwater is important. When the soil's organic matter level is depleted, it is less able to accept water, and high levels of runoff and erosion result. This means less water for plants and decreased groundwater recharge.

SOURCES

Allison, F.E. 1973. *Soil Organic Matter and its Role in Crop Production*. Scientific Publishing Co. Amsterdam, The Netherlands.

Brady, N.C., and R.R. Weil. 1999. *The Nature and Properties of Soils*. 12th ed. Macmillan Publishing Co. New York, NY.

Follett, R.F., J.W.B. Stewart, and C.V. Cole (eds). 1987. *Soil Fertility and Organic Matter as Critical Components of Production Systems*. Special Publication No.19. Soil Science Society of America. Madison, WI.

Lucas, R.E., J.B. Holtman, and J.L. Connor. 1977. Soil carbon dynamics and cropping practices. pp. 333–451. In *Agriculture and Energy* (W. Lockeretz, ed.). Academic Press.

New York, NY. See this source for the Michigan study on the relationship between soil organic matter levels and crop-yield potential.

Oshins, C. 1999. *An Introduction to Soil Health.* A slide set available at the Northeast Region SARE website: www.uvm.edu/~nesare/slide.html.

Powers, R.F., and K. Van Cleve. 1991. Long-term ecological research in temperate and boreal forest ecosystems. *Agronomy Journal* 83:11–24. This reference compares the relative amounts of carbon in soils with that in plants.

Stevenson, F.J. 1986. *Cycles of Soil: Carbon, Nitrogen, Phosphorus, Sulfur, Micronutrients.* John Wiley & Sons. New York, NY. This reference compares the amount of carbon in soils with that in plants.

Strickling, E. 1975. Crop sequences and tillage in efficient crop production. *Abstracts of the 1975 Northeast Branch American Society Agronomy Meetings.* pp. 20–29. See this source for the Maryland experiment relating soil organic matter to corn yield.

Tate, R.L., III. 1987. *Soil Organic Matter: Biological and Ecological Effects.* John Wiley & Sons. New York, NY.

5

Amount of Organic Matter in Soils

*The depletion of the soil humus supply is apt to be
a fundamental cause of lowered crop yields.*

—J.H. HILLS, C.H. JONES, AND C. CUTLER, 1908

The amount of organic matter in any particular soil is a result of a wide variety of environmental, soil, and agronomic influences. Some of these, such as climate and soil texture, are naturally occurring. Human activity also influences soil organic matter levels. Tillage, crop rotation, and manuring practices all have profound effects on the amount of soil organic matter. Pioneering work on the effect of natural influences on soil organic matter levels was carried out in the U.S. more than 50 years ago by Hans Jenny.

The amount of organic matter in soil is a result of all the additions and losses of organic matter that have occurred over the years (figure 5.1). In this chapter, we will look at why different soils have different organic matter levels. Anything that adds large amounts of organic residues to a soil may increase organic matter. On the other hand, anything that causes soil organic matter to decompose more rapidly or be lost through erosion may deplete organic matter.

If additions are greater than losses, organic matter increases. When additions are less than losses, there is a depletion of soil organic matter. When the system is in balance, and additions equal losses, the quantity of soil organic matter doesn't change over the years.

NATURAL FACTORS

Temperature

In the United States, it is easy to see how temperature affects soil organic matter levels. Traveling from north to south, average hotter temperatures lead to less soil organic matter. As the climate gets warmer, two things tend to happen

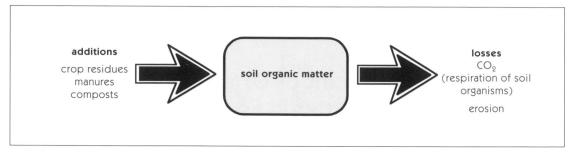

Figure 5.1 Additions and losses of organic matter from soils.

(as long as rainfall is sufficient): more vegetation is produced because the growing season is longer, and the rate of decomposition of organic materials in soils also increases, because soil organisms work more efficiently in warm weather and for longer periods of the year. This increasing decomposition with warmer temperatures becomes the dominant influence determining soil organic matter levels.

Rainfall

Soils in arid climates usually have low amounts of organic matter. In a very dry climate, such as a desert, there is little growth of vegetation. Decomposition may be very low when the soil is dry and microorganisms cannot function well. When it finally rains, a very rapid burst of decomposition of soil organic matter occurs. Soil organic matter levels generally increase as average annual precipitation increases. With more rainfall, more water is available to plants and more plant growth results. As rainfall increases, more residues return to the soil from grasses or trees. At the same time, soils in high rainfall areas may have less soil organic matter decomposition than well-aerated soils — decomposition is slowed by restricted aeration.

Soil Texture

Fine textured soils, containing high percentages of clay, tend to have naturally higher amounts of soil organic matter than coarse textured sands or sandy loams. The organic matter content of sands may be less than 1 percent; loams may have 2 to 3 percent; and clays from 4 to more than 5 percent. The strong bonds that develop between clay and organic matter seem to protect organic molecules from attack and decomposition by microorganisms. In addition, fine textured soils tend to have smaller pores and have less oxygen than coarser soils. This also causes reduced decomposition of organic matter. The lower rate of decomposition in soils with high clay contents is probably the main reason that their organic matter levels are higher than in sands and loams.

Soil Drainage and Position in the Topography

Some soils have a compact subsoil layer that doesn't allow water to drain well. Decomposition of organic matter occurs more slowly in poorly aerated soils, when oxygen is limited or absent, than in well-aerated soils. For this reason, organic matter accumulates in wet soil environments. In a totally flooded soil, one of the major structural parts of plants, lignin, doesn't decompose at all. The ultimate consequence of extremely wet or swampy conditions is the development of organic (peat or muck) soils, with organic matter contents of over 20 percent. If organic soils are artificially drained for agricultural or other uses, the soil organic matter will decompose very rapidly. When this happens, the elevation of the soil surface actually decreases. Some homeowners in Florida were fortunate to sink corner posts below the organic level. Originally level with the ground, those homes now perch on posts atop a soil surface that has decreased so dramatically the owners park under their homes.

Soils in depressions at the bottom of hills are often wet because they receive runoff, sediments (including organic matter), and seepage from up slope. Organic matter is not decomposed as rapidly in these landscape positions as in drier soils farther up slope. However, soils on a steep slope will tend to have low amounts of organic matter because the topsoil is continually eroded.

Type of Vegetation

The type of plants that grow on a soil over the years affects the soil organic matter level. The most dramatic differences are evident when soils developed under grassland are compared with those developed under forests. On natural grasslands, organic matter tends to accumulate in high amounts and to be well distributed within the soil. This is probably a result of the deep and extensive root systems of native grasses. Their roots have high "turnover" rates, for root death and decomposition constantly occurs as new roots are formed. The high levels of organic matter in soils that were once in grassland explains why these are some of the most productive soils in the world. By contrast, in forests, litter accumulates on top of the soil, and surface organic layers commonly contain over 50 percent organic matter. However, subsurface mineral layers in forest soils typically contain from less than 1 to about 2 percent organic matter.

Acidic Soil Conditions

In general, soil organic matter decomposition is slower under acidic soil conditions than at more neutral pH. In addition, acidic conditions, by inhibiting earthworm activity, encourage organic matter to accumulate at the soil surface, rather than distributed throughout the soil layers.

HUMAN INFLUENCES

Soil erosion removes topsoil rich in organic matter so that, eventually, only subsoils remain. Crop production obviously suffers when part or all of the most fertile layer of the soil is removed. Erosion is a natural process and occurs on almost all soils. Some soils naturally erode more easily than others and the problem is also greater in some regions than others. However, agricultural practices accelerate erosion. Nationwide, soil erosion causes huge economic losses. It is estimated that erosion in the United States is responsible for annual losses of $500 million in available nutrients and $18 billion in total soil nutrients.

Unless erosion is very severe, a farmer may not even realize that a problem exists, but that doesn't mean that crop yields are unaffected. In fact, yields may decrease by 5 to 10 percent when only moderate erosion occurs. Yields may

TABLE 5.1
Effects of Erosion on Soil Organic Matter and Water

Soil	Erosion	Organic Matter (%)	Available Water Capacity (%)
Corwin	slight	3.03	12.9
	moderate	2.51	9.8
	severe	1.86	6.6
Miami	slight	1.89	16.6
	moderate	1.64	11.5
	severe	1.51	4.8
Morley	slight	1.91	7.4
	moderate	1.76	6.2
	severe	1.60	3.6

—Schertz et al., 1985.

suffer a decrease of 10 to 20 percent or more with severe erosion. The results of a study of three midwestern soils, shown in table 5.1, indicate that erosion greatly influences both organic matter levels and water-holding ability. Greater amounts of erosion decreased the organic matter contents of these loamy and clayey soils. In addition, eroded soils stored less available water than soils experiencing little erosion.

Organic matter also is lost from soils when organisms decompose more organic materials during the year than are added. This occurs as a result of practices such as intensive tillage and growing crops that produce low amounts of residues (see below).

Tillage Practices

Tillage practices influence both the amount of topsoil erosion and the rate of decomposition of soil organic matter. Conventional plowing and disking of a soil to prepare a smooth seedbed breaks down natural soil aggregates and destroys large, water-conducting channels. The soil is left in a physical condition that allows both wind and water erosion.

The more a soil is disturbed by tillage practices, the greater the potential breakdown of organic matter by soil organisms. During the early years of agriculture in the United States, when colonists cleared the forests and planted crops in the East and farmers later moved to the Midwest to plow the grasslands, soil organic matter decreased rapidly. In fact, the soils were literally mined of a valuable resource — organic matter. In the Northeast and Southeast, it was quickly recognized that fertilizers and soil amendments were needed to maintain soil productivity. In the Midwest, the deep, rich soils of the tall-grass prairies were able to maintain their productivity for a long time despite accelerated soil organic matter loss and significant amounts of erosion. The reason for this was their unusually high original levels of soil organic matter.

Rapid soil organic matter decomposition by soil organisms usually occurs when a soil is worked with a moldboard plow. Incorporating residues, breaking aggregates open, and fluffing up the soil allows microorganisms to work more rapidly. It's something like opening up the air intake on a wood stove, which lets in more oxygen and causes the fire to burn hotter. In Vermont, we found a 20-percent decrease in organic matter after five years of growing corn on a clay soil that had previously been in sod for a long time. In the Midwest, 40 years of cultivation caused a 50-percent decline in soil organic matter. Rapid loss of soil organic matter occurs in the early years, because of the high initial amount of active ("dead") organic matter available to micro-organisms. After much of the active portion is lost, the rate of organic matter loss slows considerably.

With the current interest in reduced (conservation) tillage, growing row crops in the future may not have such a detrimental effect on soil organic matter. Conservation tillage practices leave more residues on the surface and cause less soil disturbance than conventional moldboard plow and disk tillage. In fact, soil organic matter levels usually increase when no-till planters place seeds in a narrow band of disturbed soil, while leaving the soil between planting rows undisturbed. The rate of decomposition of soil organic matter is lower because the soil is not drastically disturbed by plowing and disking. Residues accumulate on the surface because the soil is not inverted by plowing. Earthworm populations increase, taking some of the organic matter deeper into the soil and creating channels that help water infiltrate into the soil. Decreased erosion also results from using conservation tillage practices.

Crop Rotations and Cover Crops

At different stages in a rotation, different things may be happening. Soil organic matter may decrease, then increase, then decrease, and so forth. While annual row crops under conventional moldboard plow cultivation usually result in decreased soil organic matter, perennial legumes, grasses, or legume-grass forage crops tend to increase soil organic matter. The turnover of the roots of these hay and pasture crops, plus the lack of soil disturbance, allow organic matter to accumulate in the soil. This effect is seen in the comparison of organic matter increases when growing alfalfa compared to corn silage (figure 5.2) In addition, different types of crops result in different quantities of residues returned to the soil. When corn grain is harvested, more residues are left in the field than after soybeans, wheat, potatoes, or lettuce harvests. Harvesting the same crop in different ways leaves different amounts of residues. When corn grain is harvested, more residues remain in the field than when the entire plant is harvested for silage (figure 5.3).

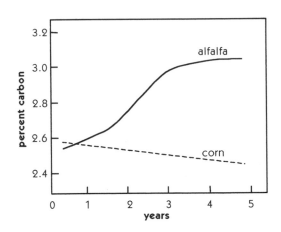

Figure 5.2 Organic carbon changes when growing corn silage or alfalfa. Redrawn from Angers, 1992.

a) corn silage

b) corn grain

Figure 5.3 Soil surface after harvest of corn silage or corn grain. Photos by Win Way.

Soil erosion is greatly reduced and topsoil rich in organic matter is conserved when rotation crops, such as grass or legume hay, are grown year-round. The extensive root systems of sod crops account for much of the reduction in erosion. Having sod crops as part of a rotation reduces loss of topsoil, decreases decomposition of residues, and builds up organic matter by the extensive residue addition of plant roots.

Use of Organic Amendments

An old practice that helps maintain or increase soil organic matter is to apply manures or other organic residues generated off the field. A study in Vermont during the 1960s and 1970s found that between 20 and 30 tons (wet weight, including straw or sawdust bedding) of dairy manure per acre were needed to maintain soil organic matter levels when silage corn was grown each year. This is equivalent to 1 to 1½ times the amount produced by a large Holstein cow over the whole year. Different manures can have very different effects on soil organic matter and nutrient availability. They differ in their initial composition and also are affected by how they are stored and handled in the field.

ORGANIC MATTER DISTRIBUTION IN SOIL

In general, more organic matter is present near the surface than deeper in the soil (see figure 5.4). This is one of the main reasons that topsoils are so productive, compared with subsoils exposed by erosion or mechanical removal of surface soil layers. Much of the plant residues that eventually become part of the soil organic matter are from the above-ground portion of plants. When the plant dies or sheds leaves or branches, it deposits residues on the surface. Although earthworms and insects help incorporate the residues on the surface deeper into the soil and the roots of some plants penetrate deeply, the highest concentrations still remain within 1 foot of the surface.

Litter layers that commonly develop on the surface of forest soils may have very high organic matter contents (figure 5.4a). Plowing forest soils after removal of the trees incorporates the litter layers into the mineral soil. The incorporated litter decomposes rapidly, and an agricultural soil derived from a light, sandy texture forest soil in the north or a silt loam in the southeast coastal plain would likely have a distribu-

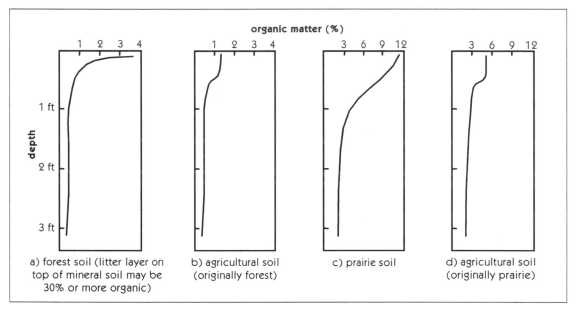

Figure 5.4 Examples of soil organic matter content with depth. Modified from Brady and Weil, 1999.

tion of organic matter similar to that indicated in figure 5.4b. Soils of the tall-grass prairies have high levels of organic matter deep into the soil profile (see figure 5.4c). After cultivation of these soils for 50 years, far less organic matter exists (figure 5.4d).

ACTIVE ORGANIC MATTER

The discussion for almost all of this chapter has been about amounts of total organic matter in soils. However, we should constantly keep in mind that we are interested in each of the different types of organic matter in soils — the living, the dead (active), and the very dead (humus). We don't just want a lot of humus in soil, we also want a lot of active organic matter to provide nutrients and aggregating glues when it is decomposed. We want the active organic matter because it supplies food to keep a di-

verse population of organisms present. As mentioned earlier, when forest or prairie soils were first cultivated, there was a drastic decrease in the organic matter content. Almost all of the decline was due to a loss of the active ("dead") part of the organic matter. It is the active fraction that increases relatively quickly when practices, such as reduced tillage, rotations, cover crops, and manures, are used to increase soil organic matter.

LIVING ORGANIC MATTER

In chapter 3, we talked about the various types of organisms that live in soils. The weight of fungi present in forest soils is much greater than the weight of bacteria. In grasslands, however, there are about equal weights of both. In agricultural soils that are routinely tilled, the weight of fungi is less than the weight of bacteria. As soils become more compact, larger pores are eliminated first. These are the pores in which soil animals, such as earthworms and beetles, live and function, so the number of such organisms in compacted soils decreases.

Different total amounts (weights) of living organisms exist in various cropping systems. In general, high populations of diverse and active soil organisms are found in systems with more complex rotations that regularly leave high amounts of crop residues and when other organic materials are added to the soil. Organic materials may include crop residues, cover crops, animal manures, and composts. Leaves and grass clippings may be an important source of organic residues for gardeners. When crops are rotated regularly, fewer parasite, disease, weed, and insect problems occur than when the same crop is grown year after year.

On the other hand, frequent cultivation reduces the number of many soil organisms as their food supplies are depleted by decomposition of organic matter. Compaction from heavy equipment causes harmful biological effects in soils. It decreases the number of medium to large pores, which reduces the volume of soil available for air, water, and populations of organisms — such as mites and springtails — that need the large spaces in which to live.

SOURCES

Angers, D.A. 1992. Changes in soil aggregation and organic carbon under corn and alfalfa. *Soil Science Society of America Journal* 56: 1244–1249.

Brady, N.C., and R.R. Weil. 1999. *The Nature and Properties of Soils*. 12th ed. Macmillan Publishing Co. New York, NY.

Carter, V.G., and T. Dale. 1974. *Topsoil and Civilization*. University of Oklahoma Press. Norman, OK.

Hass, H.J., G.E.A. Evans, and E.F. Miles. 1957. *Nitrogen and Carbon Changes in Great Plains Soils as Influenced by Cropping and Soil Treatments*. U.S. Department of Agriculture Technical Bulletin 1164. U.S. Government Printing Office. Washington, D.C. This is a reference for the large decrease in organic matter content of Midwest soils.

Jenny, H. 1980. *The Soil Resource*. Springer-Verlag. New York, NY.

Jenny, H. 1941. *Factors of Soil Formation*. McGraw-Hill. New York, NY. Jenny's early work on the natural factors influencing soil organic matter levels.

Magdoff, F.R., and J.F. Amadon. 1980. Yield trends and soil chemical changes resulting from N and manure application to continuous corn. *Agronomy Journal* 72:161–164. See this reference for further information on the studies in Vermont cited in this chapter.

National Research Council. 1989. *Alternative Agriculture*. National Academy Press. Washington, D.C.

Schertz, D.L., W.C. Moldenhauer, D.F. Franzmeier, and H.R. Sinclair, Jr. 1985. Field evaluation of the effect of soil erosion on crop productivity. pp. 9–17. In *Erosion and Soil Productivity*. Proceedings of the national symposium on erosion and soil productivity. Dec. 10–11, 1984. New Orleans, LA. American Society of Agricultural Engineers Publication 8-85. St. Joseph, MI.

Tate, R.L., III. 1987. *Soil Organic Matter: Biological and Ecological Effects*. John Wiley & Sons. New York, NY.

6

Let's Get Physical
Soil Tilth, Aeration, and Water

Moisture, warmth, and aeration; soil texture; soil fitness;
soil organisms; its tillage, drainage and irrigation;
all these are quite as important factors in the make up
and maintenance of the fertility of the soil as are manures,
fertilizers, and soil amendments.

—J.L. Hills, C.H. Jones, and C. Cutler, 1908

A soil's physical condition has a lot to do with its ability to produce crops. A degraded soil usually has reduced water infiltration and percolation (drainage into the subsoil), aeration, and root growth. This reduces the ability of the soil to supply nutrients, render harmless many hazardous compounds (such as pesticides), and to maintain a wide diversity of soil organisms. Small changes in a soil's physical conditions can have a large impact on these essential processes. Creating a good physical environment, which is a critical part of building and maintaining a healthy soil, requires a certain amount of attention and care.

Let's first look at the physical nature of a typical agricultural soil. It usually contains about 50 percent solid particles and 50 percent pores on a volume basis (figure 6.1). We discussed earlier how organic matter is only a small, but very important component of the soil. The rest of the soil particles are a mixture of various size minerals, ranging from fine-grained microscopic clay to easily visible large sand grains. The relative amounts of the various particle sizes defines the texture of a soil, such as a clay, clay loam, loam, sandy loam, or sand. Although management practices don't change this basic soil physical property, they may modify the effects of texture on other properties.

The sizes of the spaces (pores) between the particles and between aggregates are much more important than the sizes of the particles themselves. The total amount of pore space and the relative quantity of various size pores (large, medium, small, very small) govern water movement and availability for sustaining soil organisms and plants. We are interested mostly in the pores, because that's where all the important processes, such as water and air movement, take place. Soil organisms live and function in the pores, which is also where plant roots grow. Most pores in a clay loam are small (generally less than 0.0004 inch), whereas most pores in a loamy sand are large (generally still smaller than 0.1 inch). Although soil texture doesn't change over time, the total amount of pore space and the relative amount of various size pores (called the pore size distribution) are strongly affected by management.

WATER AND AERATION

The soil pore space can be filled with either water or air, and their relative amounts change as the soil wets and dries (figure 6.1). When all pores are filled with water, the soil is *saturated* and soil gases can't exchange with the atmospheric

Soil water is a rich mixture
that contains nutrients and
microorganisms.

gases. This means that carbon dioxide from respiring roots and soil organisms can't escape from the soil and oxygen can't enter, leading to undesirable *anaerobic* (no oxygen) conditions. On the other extreme, a soil with little water may have good gas exchange, but it can't supply sufficient water to plants and soil organisms.

The way in which a soil holds and releases water is pretty similar to the way it works with a sponge (figure 6.2). When it's fully saturated (you take the sponge out of a bucket of water), a sponge

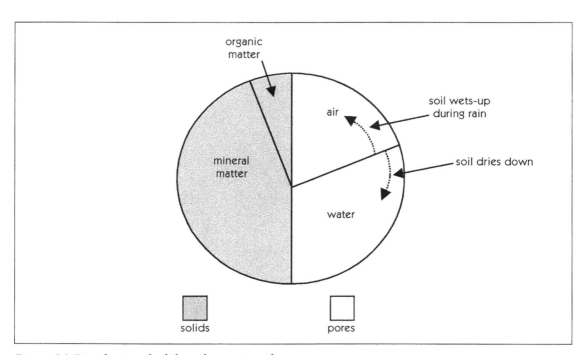

Figure 6.1 Distribution of solids and pores in soil.

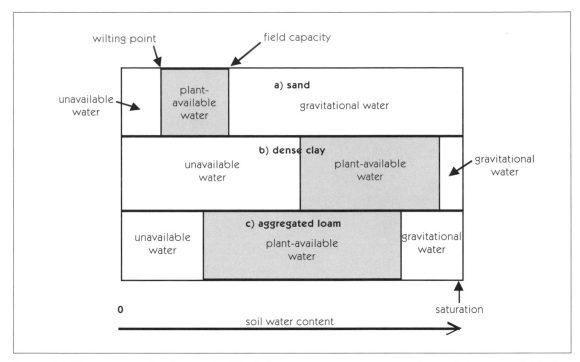

Figure 6.2 Water storage for three soils. (Shaded area represents water stored in soil that is available for plant use.)

loses water by gravity, but will stop dripping within about 30 seconds. It's only the largest pores that lose water during that rapid drainage because they are unable to hold the water against gravity. The sponge still contains a lot of water when it stops dripping. The remaining water is in the smaller pores, which hold it more tightly. The sponge's condition following drainage is almost the same as a soil reaching *field capacity* water content, which occurs after about two days of free drainage following saturation by a lot of rain or irrigation. If a soil contains mainly large pores, like a coarse sand, it loses a lot of water through gravitational drainage. This is good because these pores are now open for aeration, but it's also bad because little water remains for

plants to use, leading to frequent drought stress. Coarse sandy soils have very small amounts of water available to plants before they reach their *wilting point* (figure 6.2a). On the other hand, a dense, fine-textured soil, such as a compacted clay loam, has mainly small pores, which hold on tightly to water and don't release it as free drainage (it has little gravitational water, figure 6.2b). In this case, the soil will have long periods of poor aeration, but more plant-available water than a coarse sand. Leaching of pesticides and nitrates to groundwater is also controlled by the relative amounts of different size pores. The rapidly draining sands lose these chemicals along with the percolating water, but this is much less of a problem with clays.

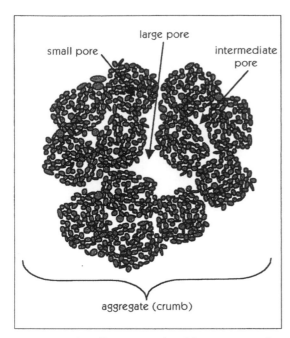

small pore

large pore

intermediate pore

aggregate (crumb)

Figure 6.3 A well aggregated soil has a range of pore sizes. This medium size soil crumb is made up of many smaller ones. Very large pores occur between the medium size aggregates.

The ideal soil is somewhere between the two extremes. This can be found in a well aggregated medium-textured loam soil (figure 6.2c, figure 6.3). Such a soil has enough large pore spaces between the aggregates to provide adequate drainage and aeration during wet periods, but also has adequate amounts of small pores and water-holding capacity to provide sufficient water to plants and soil organisms between rainfall or irrigation events. Besides holding and releasing water well, such soils also allow for good water infiltration, thereby increasing plant water availability and reducing runoff and erosion. This ideal soil condition is indicated by crumb-like aggregates, which are common in good topsoil.

GOOD SOIL TILTH IS GOOD AGGREGATION

Good aggregation, or *structure*, helps to make a high quality soil. Aggregation in the surface soil is favored by organic matter and surface residue. As pointed out earlier, a continuous supply of organic materials, roots of living plants, and mycorrhizal fungi hyphae are needed to maintain good soil aggregation. Surface residues protect the soil from wind and raindrops and moderate the temperature and moisture extremes at the soil surface. An unprotected soil may reach very high soil temperatures at the surface and become very dry. Worms and insects will move deeper into the bare soil, developing a surface zone containing few active organisms. Many small microorganisms, such as bacteria and fungi that live in thin films of water, will die or become inactive, slowing the natural process of organic matter cycling. Large and small organisms function better in a soil that is well protected by crop residue cover, a mulch, or a sod, which helps maintain good soil aggregation. An absence of both erosion and the forces that cause compaction helps maintain good surface aggregation.

WATER INFILTRATION, RUNOFF, AND EROSION

As rainfall reaches the ground, most water either infiltrates into the soil or runs off the surface (some may stand in ruts or depressions before infiltrating or evaporating). The maximum amount of rainwater that can enter a soil in a given time, called *infiltration capacity*, is influenced by the soil type, soil structure, and the soil moisture at the start of the rain. Early in a storm, water usually enters a soil readily. When the soil becomes more moist, it can soak up less water. If rain continues, runoff is produced due to the soil's reduced infiltration rate. When an intense storm hits an already saturated soil, runoff occurs very rapidly, because the soil's ability to absorb water is low. Rainfall or snowmelt on frozen ground generally poses even greater runoff concerns, as pores are blocked with ice. Runoff happens more readily with poorly structured soils, because they have fewer large pores to quickly conduct water downward. As soil tilth degrades, water infiltration decreases, producing more runoff.

Runoff water concentrates into tiny streams, which loosen soil particles and take them downhill. As runoff water gains more energy, it scours away more soil. Runoff also carries agricultural chemicals and nutrients, which end up in streams, lakes, and estuaries. Soil degradation in many of our agricultural and urban watersheds has resulted in increased runoff during intense rainfall and increased problems with flooding. Also, the lower infiltration capacity of degraded soils reduces the amount of water that is available to plants, as well as the amount that percolates through the soil into underground aquifers. This underground water feeds streams slowly. Watersheds with degraded soils experience lower stream flow during dry seasons due to low groundwater recharge and increased flooding during times of high rainfall due to high runoff.

Soil erosion is the result of exposing the soil directly to the destructive energy of raindrops and wind. Soil erosion has long been known to decrease soil quality, and at the same time cause sedimentation in downstream or downwind areas. Soil is degraded when the best soil material

— the surface layer — is removed by erosion. Erosion also selectively removes the more easily transported finer soil particles. Severely eroded soils, therefore, become low in organic matter and have less favorable physical properties, leading to a reduced ability to sustain crops and increased potential for harmful environmental impacts.

For most agricultural areas of the world today, water and wind erosion still cause extensive damage (including the spread of deserts) and remain the greatest threat to agricultural sustainability.

Some ancient farming civilizations recognized soil erosion as a problem and developed effective methods for runoff and erosion control. Evidence of ancient terracing methods are apparent in various parts of the world, notably in the Andean region of South America and in Southeast Asia. Other cultures effectively controlled erosion using mulching and intercropping, thereby protecting the soil surface from the elements. (Some ancient desert civilizations, such as the Anasazi in the Southwestern U.S. and the Nabateans in the Middle East, took advantage of surface runoff to harvest water to grow crops in downhill depressions. Their methods, however, were specific to very dry conditions.) For most agricultural areas of the world today, water and wind erosion still cause extensive damage (including the spread of deserts) and remain the greatest threat to agricultural sustainability.

Tillage degrades land even beyond promoting water and wind erosion by exposing soil to the elements. It can also cause erosion by directly moving soil down the slope to lower areas of the field. In complex topographies — such as seen in figure 6.4 — this ultimately results in the removal of surface soil from the knolls and its deposition in depressions (*swales*) at the bottom of the slopes. What causes this type of tillage erosion? Gravity causes more soil to be moved by the plow or harrow down slope than up slope. Soil is thrown farther down slope when tilling in the down slope direction than is thrown uphill when tilling in the up slope direction (figure 6.5a). Down slope tillage typically occurs at greater speed than when traveling uphill, making the situation even worse. Tillage along the

contour also results in down slope soil movement. Soil lifted by a tillage tool comes to rest at a slightly lower position on the slope (figure 6.5b). A more serious situation occurs when using a moldboard plow along the contour. This is typically performed by throwing the soil down the slope, as better inversion is obtained, than by trying to turn the furrow up the slope (figure 6.5c).

Tillage causes soil to move downhill.

Soil loss from slopes due to tillage erosion can far exceed losses from water or wind erosion. On the other hand, tillage erosion does not generally result in off-site damage, because the soil is merely moved from higher to lower

Figures 6.4 Effects of tillage erosion on soils. Photo by NRCS.

BUILDING SOILS FOR BETTER CROPS

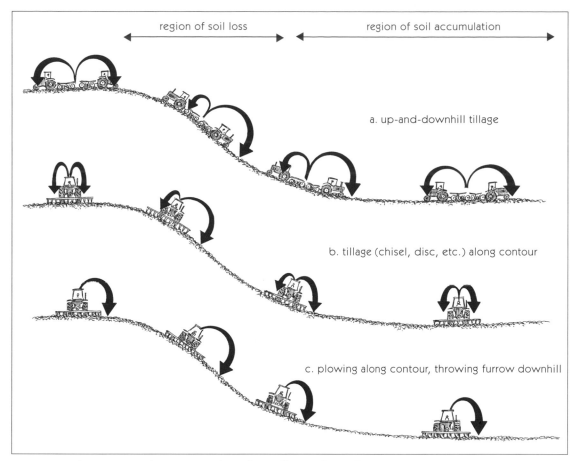

region of soil loss   region of soil accumulation

a. up-and-downhill tillage

b. tillage (chisel, disc, etc.) along contour

c. plowing along contour, throwing furrow downhill

Figure 6.5 Three causes of erosion resulting from tilling soils on slopes.

positions within a field. However, it is another reason to reduce tillage on sloping fields!

SOIL COMPACTION

A soil becomes more compact, or dense, when aggregates or individual particles of soil are forced closer together. Soil compaction has various causes and different visible effects. Three types of soil compaction may occur (figure 6.6):

- surface crusting
- plow layer compaction
- subsoil compaction

Surface crusting occurs when soil is unprotected by surface residue or a plant canopy and the energy of raindrops disperses wet aggregates, pounding them together into a thin, but dense, surface layer. The sealing of the soil reduces water infiltration and the surface forms a hard crust when dried. If the crusting occurs soon after planting, it may delay or, in some cases, prevent seedling emergence. Even when the crust is not severe enough to limit germination, it can reduce water infiltration. Soils with surface crusts are prone to high rates of runoff and erosion (see figure 4.4 in chapter 4). You can

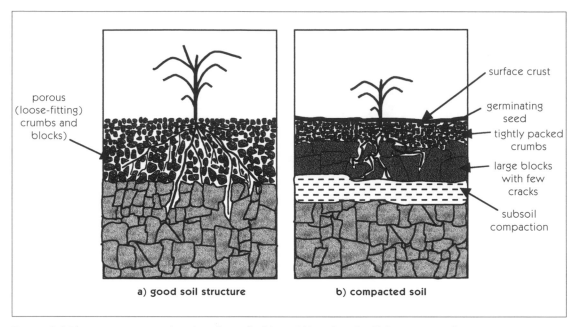

Figure 6.6 Plants growing in a) soil with good tilth and b) soil with all three types of compaction.

reduce surface crusting by leaving more residue on the surface and maintaining strong soil aggregation.

Plow layer compaction — compaction of the entire surface layer — has probably occurred to some extent in all intensively worked agricultural soils. There are three primary causes for such compaction — erosion, reduced organic matter levels, and forces exerted by field equipment. The first two result in a reduced supply of sticky binding materials and a subsequent loss of aggregation.

Compaction of soils by heavy equipment and tillage tools is especially damaging when soils are wet. To understand this, we need to know a little about soil *consistence*, or how soil reacts to external forces. At very high water content, a soil may behave like a liquid (figure 6.7) and simply flow as a result of the force of gravity —

as with mudslides during excessively wet periods. At slightly lower water contents, soil can be easily molded and is said to be *plastic*. Upon further drying, the soil will become *friable* — it will break apart rather than mold under pressure.

The point between plastic and friable soil, the *plastic limit*, has important agricultural implications. When a soil is wetter than the plastic limit, it is seriously compacted if tilled or traveled on, because soil aggregates are pushed together into a smeared, dense mass. This is why you often see smeared cloddy furrows or deep tire ruts in a field (figure 6.8). When the soil is *friable* (the water content is below the plastic limit) it breaks apart when tilled and aggregates resist compaction by field traffic. This is why the potential for compaction is so strongly influenced by timing of field operations.

BUILDING SOILS FOR BETTER CROPS

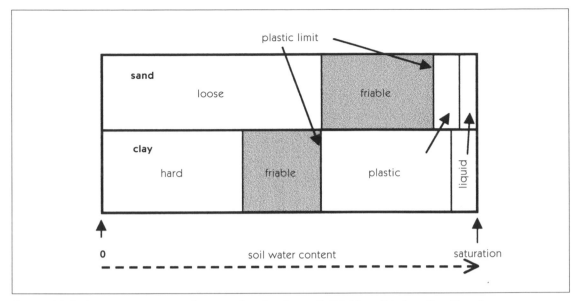

Figure 6.7 *Soil consistency states for a sand and a clay soil (friable soil is best for tillage).*

A soil's consistency is strongly affected by its texture (figure 6.7). For example, as coarse-textured sandy soils drain, they rapidly change from being plastic to friable. Fine-textured loams and clays need longer drying periods to lose enough water to become friable. This extra drying time delays field operations.

Surface crusting and plow layer compaction are especially common with intensively tilled soils. This is often part of a vicious cycle in which a compacted soil tills up very cloddy (figure 6.9a), and then requires extensive secondary tillage and packing trips to create a satisfactory seedbed (figure 6.9b). Natural aggregates break down and organic matter decomposes in the process — contributing to more compaction in the future. Although the final seedbed may be ideal at the time of planting, rainfall shortly after planting may cause surface sealing and further settling (figure 6.9c), because few sturdy

aggregates are present to prevent the soil from dispersing. The result is a dense plow layer with a crust at the surface. Some soils may hardset like cement, even after the slightest drying, slowing plant growth. Although the soil becomes softer when it re-wets, this provides only temporary relief to plants.

Figure 6.8 *Deep tire ruts in a hay field after liquid manure was applied when soil was wet and plastic.*

a) Stage 1: Cloddy soil after tillage makes for a poor seedbed.

b) Stage 2: Soil is packed and pulverized to make a fine seedbed.

c) Stage 3: Raindrops disperse soil aggregates, forming a surface crust.

Figure 6.9 Three tilth stages for a compacted soil.

Subsoil compaction — compacted soil below the normally tilled surface layer — is usually called a *plow pan*, although it's commonly caused by more than just plowing. Subsoil is easily compacted, because it is usually wetter and tends to be naturally dense, higher in clay content, lower in organic matter, and have naturally lower aggregation than topsoil. Also, subsoil is not loosened by regular tillage and cannot easily be amended with additions of organic materials, so its compaction is more difficult to manage.

Subsoil compaction is the result of either direct loading or the transfer of forces of compaction from the surface. Direct loading occurs by the pressure of a tillage implement, especially a plow or disk, pressing on the subsoil. It also occurs when a field is moldboard plowed and a set of tractor wheels is placed in the open furrow, thereby directly compacting the soil below the plow layer (figure 6.10). Subsoil compaction also occurs when farmers run heavy vehicles with poor weight distribution. The load exerted on the surface is transferred into the soil along a cone-shaped pattern (figure 6.11). With increasing depth, the force of compaction is distributed over a larger area, thereby reducing the pressure in deeper layers. When the loading force at the surface is small, say through foot

Figure 6.10 Tractor wheels in open furrow during plowing compacts subsoil.

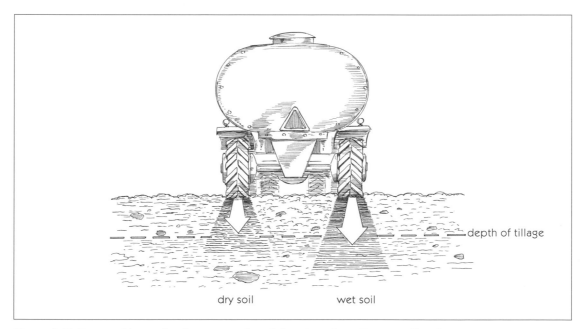

Figure 6.11 Forces of heavy loads are transferred deep into the soil, especially when wet.

traffic or a light tractor, the pressure exerted below the plow layer is minimal. But when the load is high, the pressures at depth are sufficient to cause considerable soil compaction. When the soil is wet, compaction forces near the surface are more easily transferred to the subsoil. Clearly, the most severe compaction damage to subsoils occurs by heavy vehicle traffic during wet conditions.

CONSEQUENCES OF COMPACTION

As compaction pushes soil particles closer together, the soil becomes more dense and pore space is lost. When the bulk density increases during compaction, mainly larger pores are eliminated. Loss of aggregation from compaction is particularly harmful for fine and medium-textured soils that depend on these pores for good infiltration and percolation of water, as well as air exchange with the atmosphere. Although compaction can damage coarse-textured soils, these soils depend less on aggregation, because pores between many of their particles are sufficiently large to allow good water and air movement. Compacted soil becomes hard when dried and can restrict root growth and the activity of soil organisms. The resistance to penetration, called *soil strength,* for a moist, high-quality soil is well-below the critical level (300 pounds per square inch (psi)), when root growth ceases for most crops. As the soil dries, its strength increases, but may not exceed the critical level for most (or all) of the moisture range. A compacted soil has a very narrow water content range for good root growth. It's harder in the wet range — where it may even be above the critical level,

depending on the severity of compaction. When it dries, a compacted soil hardens quicker than a well-structured soil, rapidly reaching a hardness well above the 300 psi level that restricts root growth.

Actively growing roots need pores with diameters greater than about 0.005 inch, the size of most root tips. Roots must enter the pore and anchor themselves before continuing growth. Compacted soils that have few or no large pores don't allow plants to be effectively rooted — limiting growth and water and nutrient uptake.

What happens when root growth is limited? The root system will probably have short thick roots and few fines ones or root hairs (see figure 6.6). The few existing thick roots are able to find some weak zones in the soil, often by following crooked patterns. These roots have thickened tissue and are not efficient at taking up water and nutrients. In many cases, roots in degraded soils do not grow below the tilled layer into the subsoil (see figure 6.6) — it's just too dense and hard for them to grow. Deeper root

Some Crops More Sensitive Than Others

Compaction doesn't affect all crops to the same extent. An experiment in New York found that direct-seeded cabbage and snap beans were more harmed by compaction than cucumbers, table beets, sweet corn, and transplanted cabbage. Much of the compaction damage was caused by secondary effects, such as prolonged soil saturation after rain, reduced nutrient availability or uptake, and greater pest problems.

BUILDING SOILS FOR BETTER CROPS

penetration is especially critical under rain-fed agriculture. The limitation on deep root growth by subsoil compaction increases the probability of yield losses from drought stress.

There is a more direct effect on plant growth beyond reduced root growth that limits the volume of the soil that is used for water and nutrient supply. A root system that's up against mechanical barriers sends a chemical signal to the plant shoot, which then slows down respiration and growth. This seems to be a natural survival mechanism similar to when plants experience water stress. In fact, because some of the same hormones are involved — and mechanical resistance increases when the soil dries — you usually can't separate the effects of compaction from those of drought.

THE WATER RANGE
FOR BEST PLANT GROWTH

The limitations to plant growth caused by compaction and water extremes can be combined into the concept of the *optimum water range* for plant growth — the range of water contents for which neither drought, mechanical stress, nor lack of aeration reduces plant growth (figure 6.12). This range, referred to by scientists as the *least-limiting water range*, is bounded on two sides — when the soil is too wet and when it's too dry.

The optimum water range in a well-structured soil has its *field capacity* on the wet end, as water above this water content readily drains out by gravity. On the dry end is the wilting

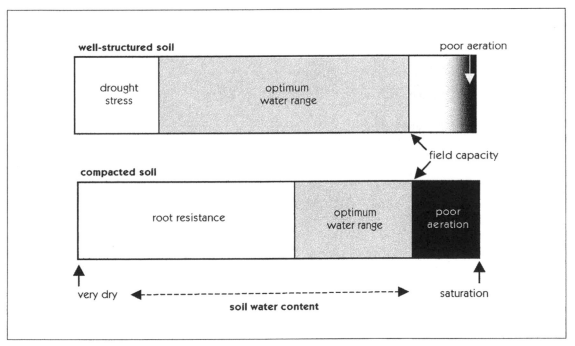

Figure 6.12 The optimum water range for crop growth for two different soils.

> Plants in compacted soils experience more stress during both wet and dry periods than plants in soils with good tilth.

point — beyond which the soil holds water too tightly to be used by plants. However, the soil water range for best growth in a compacted soil is much narrower. A severely compacted soil at field capacity is still too wet because it lacks large pores and is poorly aerated even after the soil drains. Good aeration requires about 20 percent of the pore space (about 10 percent of the volume of the whole soil) to be air-filled. On the dry end, plant growth in a compacted soil is commonly limited by soil hardness rather than by lack of available water. Plants in compacted soils experience more stress during both wet and dry periods than plants in soils with good tilth. The effects of compaction on crop yields usually depend on the length and severity of exces-sive wet or dry periods and when those periods occur relative to critical times for plant growth.

SOURCES

Hillel, D. 1991. *Out of the Earth: Civilization and the Life of the Soil.* University of California Press. Berkeley, CA.

Letey, J. 1985. Relationship between soil physical properties and crop production. *Advances in Soil Science* 1:277–294.

Ontario Ministry of Agriculture, Food, and Rural Affairs (OMFARA). 1997. *Soil Management.* Best Management Practices Series. Available from the Ontario Federation of Agriculture, Toronto, Ontario (Canada).

da Silva, A.P., B.D. Kay, and E. Perfect. 1994. Characterization of the least limiting water range of soils. *Soil Science Society of America Journal* 58:1775–1781.

Unger, P.W., and T.C. Kaspar. 1994. Soil compaction and root growth: a review. *Agronomy Journal* 86:759–766.

Soehue, W. 1958. Fundamentals of pressure distribution and soil compaction under tractor tires. *Agricultural Engineering* 39:276–290.

7

Nutrient Cycles and Flows

*Increasingly…emphasis is being laid on the
direction of natural forces, on the conservation
of inherent richness, on the acquirement of plant
food supplies from the air and subsoil.*

—J.L. HILLS, C.H. JONES, AND C. CUTLER, 1908

We used the term *cycle* earlier when discussing the flow of nutrients from soil to plant to animal to soil, as well as global carbon and nitrogen cycles (chapter 4). Some farmers depend more on natural soil nutrient cycles — as contrasted with purchased commercial fertilizers — to provide fertility to plants. Is it really possible to depend forever on the natural cycling of all the nutrients the crop needs? Let's first consider what a cycle really is and how it differs from the other ways that nutrients move from one location to another.

When nutrients move from one place to another, that is a *flow*. There are many different types of nutrient flows that can occur. When you buy fertilizers or animal feeds, nutrients are "flowing" onto the farm. When you sell sweet corn, apples, alfalfa hay, or milk, nutrients are "flowing" off the farm. Flows that involve products entering or leaving the farm gate are managed intentionally, whether or not you are thinking about nutrients. Other flows are unplanned: when nitrate is lost from the soil by leaching to groundwater or when runoff waters take nutrients along with eroded topsoil to a nearby stream.

When crops are harvested and brought to the barn to feed animals, that is a nutrient flow, as is the return of animal manure to the land. Together these two flows are a true cycle, because nutrients return to the fields from which they came. In forests and natural grassland, the cycling of nutrients is very efficient. In the early stages of agriculture, where almost all people lived near their fields, nutrient cycling was also efficient (figure 7.1a). However, in many types of agriculture, especially modern "industrial-style" farming, there is little real cycling of nutrients, because there is no easy way to return nutrients shipped off the farm. In addition, nutrients in crop residues don't cycle very efficiently when the soil is without living plants for long periods, and nutrient runoff and leaching losses are much larger than from natural systems.

The first major break in the cycling of nutrients occurred as cities developed and nutrients began to routinely travel with the farm products to feed the growing urban populations. Few nutrients now return to the soils that grew them many miles away (figure 7.1b, 7.1c). The accumulated nutrients in urban sewage have polluted waterways around the world. Even with the building of many new sewage treatment plants in the 1970s and 1980s, effluent containing nutrients still flows into waterways, and sewage sludges are not always handled in an environmentally sound manner.

The trend to farm specialization has resulted in the second break in nutrient cycling by separating animals from the land that grows their feed. With specialized animal facilities (figure 7.1c), nutrients accumulate in manure at the same time that crop farmers purchase large quantities of fertilizers to keep their fields from becoming nutrient deficient.

DIFFERING FLOW PATTERNS

Different types of farms may have distinctly different nutrient flow patterns. Farms that are exclusively growing grain or vegetables have a relatively high annual nutrient export (figure 7.2a). Nutrients usually enter the farm as either commercial fertilizers or various amendments and leave the farm as plant products. Some cycling of nutrients occurs as crop residues are returned to the soil and decompose. A large

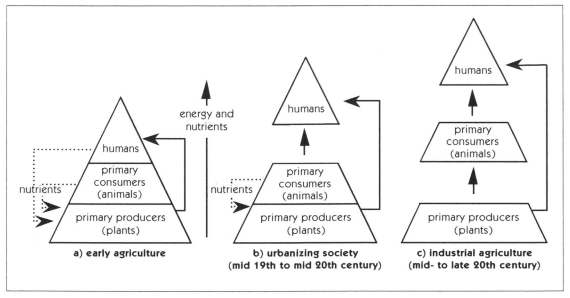

Figure 7.1 The patterns of nutrient flows change over time. From Magdoff et al., 1997.

BUILDING SOILS FOR BETTER CROPS

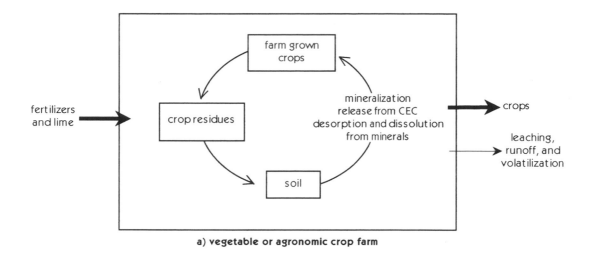

a) vegetable or agronomic crop farm

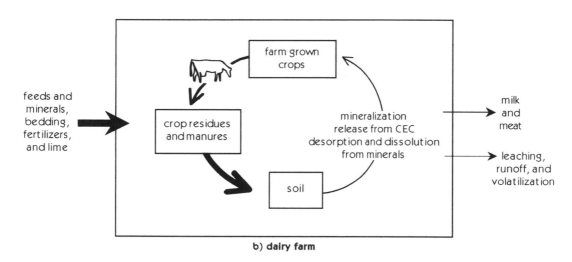

b) dairy farm

Figure 7.2 Nutrient flows and cycles on a) crop and b) dairy farms (larger flows indicated by thicker lines).

nutrient outflow is common, however, because a large portion of the crop is usually exported off the farm. For example, an acre of a good crop of tomatoes or onions usually contains over 100 lbs. of nitrogen, 20 lbs. of phosphorus, and 100 lbs. of potassium. For agronomic crops, the annual exports of nutrients is about 100 lbs. of nitrogen, 6 lbs. of phosphorus, and 50 lbs. of potassium per acre for corn grain and about 150 lbs. of nitrogen, 20 lbs. of phosphorus and 130 lbs. of potassium per acre for grass hay.

It should be fairly easy to balance inflows and outflows on crop farms, at least theoretically. In practice, under good management, nutrients are depleted a bit by crop growth and removal until soil test levels fall too low, and then they're raised again with fertilizers or manures (see chapter 19).

A contrasting situation occurs on dairy farms, if all of the forage is produced on the farm, but grains and minerals are purchased (figure 7.2b). In this situation, there are more sources for the nutrients coming onto the farm — with feeds and minerals for animal consumption usually a larger source than fertilizers. Most of the nutrients consumed by animals end up in the manure — 60 to over 90 percent of the nitrogen, phosphorus, and potassium. Compared with crop farms, more nutrients flow onto many dairy farms and fewer flow off per acre. Under this situation, nutrients will accumulate on the farm and may eventually cause environmental harm from excess nitrogen or phosphorus.

Two different nutrient flows occur when manure on livestock farms is applied to the fields used for growing the feeds. The nutrients in the manure that came from farm-grown feed sources are completing a true cycle. The nutrients in the manure that entered the farm as purchased feeds and mineral supplements are not participating in a true cycle. These nutrients are completing a flow that might have started in a far-away farm or mine and are now just being transported from the barn to the field.

Animal operations that import all feeds and that have a limited land base to use the manure have the greatest potential to accumulate high amounts of nutrients. Contract growers of chickens are an example of this practice.

If there is enough cropland to grow most of the grain and forage needs, low amounts of imported nutrients and export per acre will result. The relatively low amounts of nutrients exported per acre from animal products makes it easier to rely on nutrient cycling on a mixed livestock-crop farm that produces most of its feed, than on a farm growing only crops.

IMPLICATIONS OF NUTRIENT FLOW PATTERNS

Long distance transportation of nutrients is central to the way in which the modern food system functions. On average, the food we eat has traveled about 1,300 miles from field to processor to distributor to consumer. Exporting wheat from the U.S. Pacific Northwest to China involves an even longer distance, as does import of apples from New Zealand to New York. The nutrients in concentrated commercial fertilizers also travel large distances from the mine or factory to distributors to the field. The specialization of the corn and soybean farms of the Midwest and the hog and chicken mega-farms centralized in a few regions, such as Arkansas, the Delmarva Peninsula and North Carolina, has created a unique situation. The long distance flows of nutrients from crop farms to animal farms requires the purchase of fertilizers on the crop farms; meanwhile, the animal farms are overloaded with nutrients.

Of course, the very purpose of agriculture in the modern world — the growing of food and fiber and the use of the products by people living away from the farm — results in a loss of nutrients from the soil, even under the best possible management. In addition, leaching losses of nutrients, such as calcium, magnesium, and potassium, are accelerated by natural acidification, as well as by acidification caused by the use of fertilizers. Soil minerals — especially in the "young" soils of glaciated regions and in arid regions not subject to much leaching — may supply lots of phosphorus, potassium, calcium, and magnesium and many other nutrients. A soil with plentiful active organic matter also may supply nutrients for a long time. Eventually, however, nutrients will need to be applied to a continually cropped soil. Nitrogen is the only nutrient you can "produce" on the farm — legumes and their bacteria working together can remove nitrogen from the atmosphere and change it into forms that plants can use. However, sooner or later you will need to apply some phosphorus or potassium, even to the richest soils. If the farm is in a mixed crop-livestock system that exports only animal products, it may take a very long time to deplete a rich soil, because so few nutrients per acre are exported with those products. For crop farms, especially in humid regions, the depletion occurs more rapidly, because more nutrients are exported per acre each year.

On average, the food we eat has traveled about 1,300 miles from field to processor to distributor to consumer.

The issue eventually becomes not whether nutrients will be imported onto the farm, but rather, what source of nutrients you should use. Will the nutrients brought onto the farm be commercial fertilizers, traditional amendments (limestone), biologically fixed nitrogen, imported feeds or minerals for livestock, organic materials, such as manures, composts and sludges, or some combination of sources?

Three Different Flow Patterns

There are three main nutrient flow patterns, with each one having implications for the long-term functioning of the farm. Imports of nutrients may be less than exports, imports may be greater than exports, and imports may equal exports.

Imports are less than exports. For farms "living off capital" and drawing down the supplies of nutrients from minerals and organic matter, nutrient concentrations continually decline. This can continue for awhile, just like a person can continue to live off savings in a bank account until the money runs out. At some point, the availability of one or more nutrients becomes so low that crop yields decrease. If this condition is not remedied, the farm becomes less and less able to produce food and its economic condition will decline. This is clearly not a desirable situation for either the farm or the country. Unfortunately, the low productivity of much of Africa's agricultural lands is partially caused by this type of nutrient flow pattern.

Imports are much larger than exports. Animal farms with inadequate land bases pose a different type of problem. As animal numbers increase, relative to the available cropland and pasture, larger purchases of feeds (containing nutrients) are necessary. As this occurs, there is less land available — relative to the nutrient loads — to spread manure. Ultimately, the operation exceeds the capacity of the land to assimilate all the nutrients and pollution of ground and surface waters occurs. This pattern of nutrient flow is not environmentally acceptable. However, under current conditions, it may be more economical than a more balanced pattern.

Imports and exports are close to balance.
From the environmental perspective and for the sake of long-term soil health, fertility should be raised to — and then maintained at — optimal levels. The best way to keep desirable levels once they are reached is to roughly balance inflows and outflows. Soil tests can be very helpful to fine-tune a fertility program and make sure that levels are not building up too high or being drawn down too low (see chapter 19). This can be a challenge and may not be economically possible for all farms. This is easier to do on a mixed crop-livestock farm than on either a crop farm or a livestock farm that depends significantly on imported feeds.

SOURCES

Magdoff, F., L. Lanyon, and W. Liebhardt. 1997. Nutrient cycling, transformations, and flows: Implications for a more sustainable agriculture. *Advances in Agronomy* 60: 1–73.

Magdoff, F., L. Lanyon, and W. Liebhardt. 1998. *Sustainable Nutrient Management: A Role for Everyone*. Northeast Region Sustainable Agriculture Research and Education Program. Burlington, VT.

PART TWO

Ecological Soil and Crop Management

8

Managing for High Quality Soils:

Organic Matter, Soil Physical Condition, Nutrient Availability

*Because organic matter is lost from the soil through decay,
washing, and leaching, and because large amounts are
required every year for crop production, the necessity of
maintaining the active organic-matter content of the soil, to
say nothing of the desirability of increasing it on many
depleted soils, is a difficult problem.*

—A. F. GUSTAFSON, 1941

Building high-quality soils takes a lot of thought and action over many years. Of course, there are things that can be done right off — plant a cover crop this fall or just make a New Year's resolution not to work soils that really aren't ready in the spring (and then stick with it). Other changes take more time. You need to study carefully before drastically changing crop rotations, for example. How will the new crops be marketed and are the necessary labor and machinery available?

There are three different general management approaches to enhancing soil health. First, various practices to build up and maintain high levels of soil organic matter are key. Second, developing and maintaining the best possible soil physical condition often requires other types of practices, in addition to those that directly impact soil organic matter. Paying better attention to soil tilth and compaction is more important than ever, because of the use of very heavy field machinery. Lastly, although good organic matter management goes a long way toward providing good plant nutrition in an environmentally sound way, good nutrient management involves additional practices.

ORGANIC MATTER MANAGEMENT

It is difficult to be sure exactly why problems develop when organic matter is depleted in a particular soil. However, even in the early 20th century, agricultural scientists proclaimed, "Whatever the cause of soil unthriftiness, there is no dispute as to the remedial measures. Doctors may disagree as to what causes the disease, but agree as to the medicine. Crop rotation! The use of barnyard and green manuring! Humus

63

maintenance! These are the fundamental needs" (Hills, Jones, and Cutler, 1908). Close to a century later, these are still some of the major remedies available to us.

There seems to be a contradiction in our view of soil organic matter. On one hand, we want crop residues, dead microorganisms, and manures to decompose. If soil organic matter doesn't decompose, then no nutrients are made available to plants, no glue to bind particles is manufactured, and no humus is produced to hold on to plant nutrients as water leaches through the soil. On the other hand, numerous problems develop when soil organic matter is significantly depleted through decomposition. This dilemma of wanting organic matter to decompose, but not wanting to lose too much, means that organic materials must be continually added to the soil. A supply of active organic matter must be maintained so that humus can continually accumulate. This does not mean that organic materials must be added to each field every year. However, it does mean that a field cannot go without additions of organic residues for many years without paying the consequences.

Do you remember that plowing a soil is similar to opening up the air intake on a wood stove? What we really want in soil is a slow, steady burn of the organic matter. You get that in a wood stove by adding wood every so often and making sure the air intake is on a medium setting. In soil, you get a steady burn by adding organic residues regularly and by not disturbing the soil too often.

There are three general management strategies for organic matter management. First, use crop residues more effectively and find new sources of residues to add to soils. New residues can include those you grow on the farm, such as cover crops, or those available from various local sources. Second, be sure to use a number of different types of materials — crop residues, manures, composts, cover crops, leaves, etc. It is important to provide varied residue sources to help develop and maintain a diverse group of soil organisms. Third, implement practices that decrease the loss of organic matter from soils because of accelerated decomposition or erosion.

Soil Organic Matter Levels

Raising and maintaining soil organic matter levels. It is not easy to dramatically increase the organic matter content of soils or even to maintain good levels once they are reached. Improving organic matter content requires a sustained effort that includes a number of approaches to return organic materials to soils and minimize soil organic matter losses. It is especially difficult to raise the organic matter content of soils that are very well aerated, such as coarse sands, because added materials are decomposed so rapidly. Soil organic matter levels can be maintained with less organic residue in high clay-content soils with restricted aeration than in coarse-textured soils.

All practices that help to build organic matter levels do at least one of two things — add more organic materials than was done in the past or decrease the rate of organic matter loss from

---

Soil Organic Matter Management Strategies

✓ Increase additions of organic residues to soils.
✓ Use varied sources of organic materials.
✓ Decrease losses of organic matter from soils.

BUILDING SOILS FOR BETTER CROPS

soils (table 8.1). Those practices that do both may be especially useful. Practices that reduce losses of organic matter either slow down the rate of decomposition or decrease the amount of erosion. Soil erosion must be controlled to keep organic matter-enriched topsoil in place. In addition, organic matter added to a soil must either match or exceed the rate of loss by decomposition. These additions can come from manures and composts brought from off the field, crop residues and mulches remaining following harvest, or cover crops. Reduced tillage lessens the rate of organic matter decomposition and also may result in less erosion. When reduced tillage increases crop growth and residues returned to soil, it is usually a result of better water infiltration and storage and less surface evaporation. It is not possible in this book to give specific soil organic matter management recommendations for all situations. In chapters 9 through 15, we will evaluate management options and issues associated with their use.

How much organic matter is enough? Unlike the case with plant nutrients or pH levels, there are no accepted guidelines for organic matter content. We do know some general guidelines. For example, 2 percent organic matter in a sandy soil is very good, but in a clay soil, 2 percent indicates a greatly depleted situation. The complexity of soil organic matter composition, including biological diversity of organisms as well as the actual organic chemicals present, means that there is no simple interpretation for total soil organic matter tests.

Using Organic Materials

Crop residues. Crop residues are usually the largest source of organic materials available to farmers. The amount of crop residue left after harvest varies depending on the crop. Soybeans, potatoes, lettuce and corn silage leave little residue. Small grains, on the other hand, leave more residue, while sorghum and corn harvested for

TABLE 8.1
Effects of Different Management Practices on Gains and Losses of Organic Matter

Management Practice	Gains Increases	Losses Decreases
■ Add materials from off the field (manures, composts, other organic materials)	yes	no
■ Better utilize crop residue	yes	no
■ Include high residue producing crops in rotation	yes	no
■ Include sod crops (grass/legume forages) in rotation	yes	yes
■ Grow cover crops	yes	yes
■ Reduce tillage intensity	yes/no*	yes
■ Use conservation practices to reduce erosion	yes/no*	yes

* practice may increase crop yields, resulting in more residue

grain leave the most. A ton or more of crop residues per acre may sound like a lot of organic material being returned to the soil. However, keep in mind that after residues are decomposed by soil organisms only about 10 to 20 percent of the original amount is converted into stable humus.

The amount of roots remaining after harvest also can range from very low to fairly high. For a crop of corn, roots may account for over a ton of dry weight per acre (thus more than 4½ tons of surface residues plus roots — about 60 percent of the total plant — remain following a Midwest grain harvest of about 120 bu. per acre).

The estimated root residues (from Prince Edward Island in Canada) give some idea of the differences that you might find (table 8.2).

Some farmers remove above ground residues from the field for use as animal bedding or to make compost. Later, these residues return to contribute to soil fertility as manures or composts. Sometimes, residues are removed from fields, to be used by other farmers or to make another product. There is renewed interest in using crop residues as a wood substitute to make a variety of products, such as particleboard. This activity could cause considerable harm because residues are not returned to soils.

Crop Residues

The amount of residue left in the field after harvest depends on the type of crop and its yield. The table on the left contains the amounts of residues found in California's highly productive, irrigated San Joaquin Valley. These residue amounts are higher than would be found on most farms, but the relative amounts for the various crops are interesting.

Crop Residues in the San Joaquin Valley (California)

CROP	TONS/ACRE
Corn (grain)	5
Broccoli	3
Cotton	2½
Wheat (grain)	2½
Sugarbeets	2
Safflower	1½
Tomatoes	1½
Lettuce	1
Corn (silage)	½
Garlic	½
Wheat (after baling)	¼
Onions	¼

—MITCHELL ET AL., 1999

Residues of Common Crops in the Midwest and Great Plains

CROP	TONS/ACRE
Corn (120 bu.)	3½
Sorghum (80 bu.)	2½
Wheat (35 bu.)	2
Soybeans (35 bu.)	less than 1

—FROM VARIOUS SOURCES

TABLE 8.2
Estimated Root Residue Produced by Crops

Crop	Estimated Root Residues (lbs./acre)
Italian ryegrass	2,600–4,500
Winter cereal	2,200–2,600
Red clover	2,200–2,600
Spring cereal	1,300–1,800
Soybeans	500–900
Potatoes	300–600

—Topp et al., 1995

Burning of wheat, rice, and other crop residues in the field is a common practice in parts of the United States as well as in other countries. Residue is usually burned to help control insects or diseases or to make next year's fieldwork easier. Residue burning may be so widespread in a given area that it causes a local air pollution problem. Burning also diminishes the amount of organic matter returned to the soil and the amount of protection against raindrop impact.

Sometimes, important needs for crop residues and manures may prevent their use in maintaining or building soil organic matter. For example, straw may be removed from a grain field to serve as mulch in a strawberry field. These trade-offs of organic materials can sometimes cause a severe soil-fertility problem if allowed to continue for a long time. This issue is of much more widespread importance in developing countries where resources are scarce. There, crop residues and manures frequently serve as fuel for cooking or heating when gas, coal, oil, or wood are not available. In addition, straw may be used in making bricks or used as thatch for housing or to make fences. Although it is completely understandable that people in resource-poor regions use residues for such purposes, the negative effects of these uses on soil productivity can be substantial. An important way to increase agricultural productivity in developing countries is to find alternative sources for fuel and building materials to replace the crop residues and manures traditionally used.

Using residues as mulches. Crop residues or composts can be used as a mulch on the soil surface. This occurs routinely in some reduced tillage systems when high residue-yielding crops are grown or when killed cover crops remain on the surface. In some small-scale vegetable and berry farming, mulching is done by applying straw from off-site. Strawberries grown in the colder northern parts of the country are routinely mulched with straw for protection from winter heaving. The straw is blown on in late fall and is then moved into the interrows in the spring, providing a surface mulch during the growing season.

Mulching has numerous benefits, including:

- enhanced water availability to crops (better infiltration into the soil and less evaporation from the soil);
- weed control;
- less extreme changes in soil temperature;
- reduced splashing of soil onto leaves and fruits and vegetables (making them look better as well as reducing diseases); and
- reduced infestations of certain pests (Colorado potato beetle on potatoes is less severe when potatoes are grown in a mulch system).

On the other hand, residue mulches in cold climates can delay soil warming in the spring, reduce early season growth, and increase problems with slugs during wet periods. Of course, one of the reasons for the use of plastic mulches (clear and black) for crops like tomatoes and melons is to help warm the soil.

Effects of Residue Characteristics on Soil

Decomposition rates and effects on aggregation. Residues of various crops and manures have different properties and, therefore, have different effects on soil organic matter. Materials with low amounts of hard-to-degrade hemicellulose and lignin, such as cover crops when still very green and soybean residue, decompose rapidly (figure 8.1) and have a shorter-term effect on soil organic matter levels than residues with high levels of these chemicals (for example, corn and wheat). Manures, especially those that

contain lots of bedding (high in hemicellulose and lignin), are decomposed more slowly and tend to have more long-lasting effects on total soil organic matter than crop residues or manures without bedding. Also, cows — because they eat a diet containing lots of forages, which they do not completely decompose — have manure with longer lasting effects on soils than non-ruminants, such as chickens and hogs, that are fed exclusively a high-grain/low-fiber diet. Composts contribute little active organic matter to soils, but add a lot of well decomposed materials (figure 8.1).

In general, residues containing a lot of cellulose and other easy-to-decompose materials will have a greater effect on soil aggregation than compost, which has already undergone decomposition. Because aggregates are formed from by-products of decomposition by soil organisms, organic additions like manures, cover crops, and

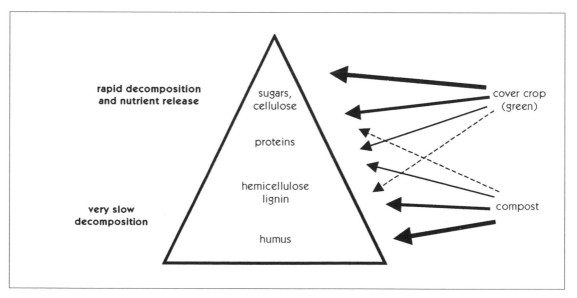

Figure 8.1 Different types of residues have varying effects on soils (thicker lines indicate more material, dashed line indicates very small percent of that type). Modified from Oshins, 1999.

BUILDING SOILS FOR BETTER CROPS

straw will enhance aggregation more than compost. (However, adding compost does improve soils in many ways, including increasing the water holding capacity.)

Although it's important to have adequate amounts of organic matter in soil, that isn't enough. A variety of residues is needed to provide food to a diverse population of organisms, nutrients to plants, and to furnish materials that promote aggregation. Residues low in hemicellulose and lignin usually have very high levels of plant nutrients. On the other hand, straw or sawdust (containing a lot of lignin) can be used to build up organic matter, but a severe nitrogen deficiency and an imbalance in soil microbial populations will occur unless a readily available source of nitrogen is added at the same time (see discussion of C:N ratios below). In addition, when insufficient N is present, less of the organic material added to soils actually ends up as humus.

C:N Ratio of organic materials and nitrogen availability. The ratio of the amount of a residue's carbon to the amount of nitrogen influences nutrient availability and the rate of decomposition. The ratio, usually referred to as the C:N ratio, may vary from around 15:1 for young plants, to between 50 to 80:1 for the old straw of crop plants, to over 100:1 for sawdust. For comparison, the C:N ratio of soil organic matter is usually in the range of about 10 to 12:1 and the C:N of soil microorganisms is around 7:1.

The C:N ratio of residues is really just another way of looking at the percentage of nitrogen (figure 8.2). A high C:N residue has a low percentage of nitrogen. Low C:N residues have relatively high percentages of nitrogen. Crop residues are usually pretty close to 40 to 45 percent carbon, and this figure doesn't change much from plant to plant. On the other hand, nitrogen content varies greatly depending on the type of plant and its stage of growth.

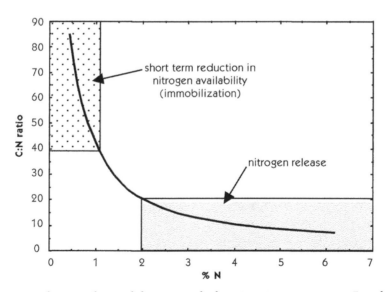

Figure 8.2 Nitrogen release and immobilization with changing nitrogen content. Based on data of Vigil and Kissel, 1991.

If you want crops growing immediately following the application of organic materials, care must be taken to make nitrogen available.

Nitrogen availability from residues varies considerably. Some residues, such as fresh, young, and very green plants, decompose rapidly in the soil and, in the process, may readily release plant nutrients. This could be compared to the effect of sugar eaten by humans, which results in a quick burst of energy. Some of the substances in older plants and in the woody portion of trees, such as lignin, decompose very slowly in soils. Materials, such as sawdust and straw, mentioned above, contain little nitrogen. Well-composted organic residues also decompose slowly in the soil because they are fairly stable, having already undergone a significant amount of decomposition.

Mature plant stalks and sawdust that have C:N over 40:1 (table 8.3) may cause temporary problems for plants. Microorganisms using materials containing 1 percent nitrogen (or less) need extra nitrogen for their growth and reproduction. They will take the needed nitrogen from the surrounding soil, diminishing the amount of nitrate and ammonium available for crop use. This reduction of soil nitrate and ammonium by microorganisms decomposing high C:N residues is called *immobilization* of nitrogen.

When microorganisms and plants compete for scarce nutrients, the microorganisms usually win, because they are so well distributed in the soil. Plant roots are in contact with only 1 to 2 percent of the entire soil volume whereas microorganisms populate almost the entire soil. The length of time during which the nitrogen nutrition of plants is adversely affected by immobilization depends on the quantity of residues applied, their C:N ratio, and other factors influencing microorganisms, such as fertilization practices, temperature, and moisture conditions. If the C:N ratio of residues is in the teens or low 20s, corresponding to greater than 2 percent nitrogen, then there is more nitrogen present than the microorganisms need for residue decomposition. When this happens, extra nitrogen becomes available to plants fairly quickly. Green manure crops and animal manures are in this group of residues. Residues with C:N in the mid-20s to low 30s, corresponding to about 1 to 2 percent nitrogen, will not have much effect on short-term nitrogen immobilization or release.

Sewage sludge on your fields? In theory, the use of sewage sludges on agricultural lands makes sense as a way to resolve problems related to people living in cities, far removed from the land that grows their food. However, there are some troublesome issues associated with ag-

TABLE 8.3
C:N Ratios of Selected Organic Materials

MATERIAL	C:N
Soil	10–12
Poultry manure	10
Clover and alfalfa (early)	13
Compost	15
Dairy manure (low bedding)	17
Alfalfa hay	20
Green rye	36
Corn stover	60
Wheat, oat, or rye straw	80
Oak leaves	90
Fresh sawdust	400
Newspaper	600

—C:N VALUES FROM VARIOUS SOURCES

C:N Ratio of Active Organic Matter

As residues are decomposed by soil organisms, carbon is lost as CO_2 while nitrogen is mostly conserved. This causes the C:N ratio of decomposing residues to decrease. Although the C:N ratio for most agricultural soils is in the range of 10 to 12:1, the different types of organic matter within a soil have different C:N ratios. The larger particles of soil organic matter have higher C:N ratios, indicating that they are less decomposed than smaller fractions. Microscopic evidence also indicates that the larger fractions are less decomposed than the smaller particles.

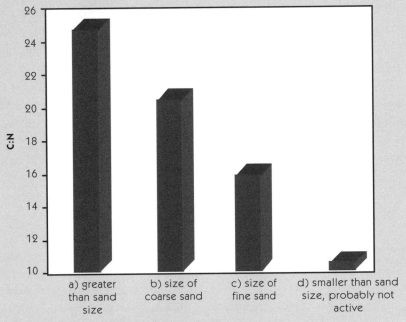

Figure 8.3 C:N ratio of different size fractions of organic matter.

—MAGDOFF, F., UNPUBLISHED DATA, AVERAGE FOR THREE SOILS.

ricultural use of sludges. By far, the most important problem is that they frequently contain contaminants from industry and from various products used around the home. Although many of these metal contaminants naturally occur at low levels in soils and plants, their high concentrations in some sludges create a potential hazard. The U.S. standards for toxic materials in sludges are much more lenient than those in other industrialized countries and they permit higher loading of potentially toxic metals. So, although you are allowed to use many sludges, you should carefully examine a sludge's contents before applying it to your land.

Another issue is that sludges are produced by varied processes and, therefore, have different properties. Most sludges are around neutral pH, but, when added to soils, cause some degree of acidification, as do most nitrogen fertilizers. Because many of the problem metals are more soluble under acidic conditions, the pH of soils receiving these materials should be monitored and maintained at around 6.8 or above. On the other hand, lime (calcium hydroxide and ground limestone are used together) is added to some sludges to raise the pH and kill disease bacteria. The resulting "lime-stabilized" sludge has extremely high levels of calcium, relative to potassium and magnesium. This type of sludge should be used primarily as a liming source and levels of magnesium and potassium in the soil need to be carefully monitored to be sure they are present in reasonable amounts, compared with the high levels of added calcium.

The use of "clean" sludges — those containing low levels of metal and organic contaminants — for agronomic crops is certainly an acceptable practice. Sludges should not be applied to soils when growing crops for direct human consumption, unless it can be demonstrated that, in addition to low levels of potentially toxic materials, organisms dangerous to humans are absent.

Application rates for organic materials. The amount of residue added to a soil is often determined by the cropping system. The crop residues can be left on the surface or incorporated by tillage. Different amounts of residue will remain under different crops, rotations, or harvest practices. For example, three or more tons per acre of leaf, stalk, and cob residues remain in the field when corn is harvested for grain. If the entire plant is harvested to make silage, there is little left except the roots.

When "imported" organic materials are brought to the field, you need to decide how much and when to apply them. In general, application rates of these residues will be based on their probable contribution to the nitrogen nutrition of plants. We don't want to apply too much available nitrogen because it will be wasted. Nitrate from excessive applications of organic sources of fertility may leach into groundwater just as easily as nitrate originating from purchased synthetic fertilizers. In addition, excess nitrate in plants may cause health problems for humans and animals.

Sometimes the fertility contribution of phosphorus may be the main factor governing application rates of organic material. Excess phosphorus entering lakes can cause an increase in the growth of algae and other aquatic weeds, decreasing water quality for drinking and recreation. In these locations, farmers must be careful to avoid loading the soil with too much phosphorus, from either commercial fertilizers or organic sources.

Effects of residue and manure accumulations. When any organic material is added to soil, it decomposes relatively rapidly at first. Later, when only resistant parts (for example, straw stems high in lignin) are left, the rate of decomposition decreases greatly. This means that although nutrient availability diminishes each year after adding a residue to the soil, there are still long-term benefits from adding organic materials. This can be expressed by using a "decay series." For example, 50, 15, 5, and 2 percent of the amount of nitrogen added in manure may be released in the first, second, third, and fourth years following addition to soils. In other words, crops in a regularly manured field get some nitrogen from manure that was applied in past years. So, if you are starting to manure a field, somewhat more manure will be needed in the first year than will be needed in years 2, 3, and 4 to supply the same total amount of nitrogen to a crop each year. After some years, you may need only half of the amount used to supply all the nitrogen needs in the first year.

Organic Matter Management on Different Types of Farms

Animal-based farms. It is certainly easier to maintain soil organic matter in animal-based agricultural systems. Manure is a valuable by-product of having animals. Animals also can use sod-type grasses and legumes as pasture, hay, and haylage (hay stored under air-tight conditions so that some fermentation occurs). It is easier to justify putting land into perennial forage crops for part of a rotation when there is an economic use for the crops. Animals need not be on the farm to have positive effects on soil fertility. A farmer may grow hay to sell to a neighbor and trade for some animal manure from the neighbor's farm, for example. Occasionally, formal agreements between dairy farmers and vegetable growers lead to cooperation on crop rotations and manure application.

Systems without animals. It is more challenging, although not impossible, to maintain or increase soil organic matter on non-livestock farms. It can be done by using reduced tillage, cover crops, intercropping, living mulches, rotations that include crops with high amounts of residue left after harvest, and attention to other erosion-control practices. Organic residues, such as leaves or clean sewage sludges, can sometimes be obtained from nearby cities and towns. Straw

---

Maintaining Organic Matter in Small Gardens

There are a number of different ways that home gardeners can maintain soil organic matter. One of the easiest is using lawn grass clippings for mulch during the growing season. The mulch can then be worked into the soil or left on the surface to decompose until the next spring. Also, leaves can be raked up in the fall and applied to the garden. Cover crops can also be used on small size gardens. Of course, manures, composts, or mulch straw can also be purchased.

There are a growing number of small-scale market gardeners, many with insufficient land to rotate into a sod type crop. They also may have crops in the ground late into the fall, making cover cropping a challenge. One possibility is to establish cover crops by over-seeding after the last crop of the year is well established. Another source of organic materials — grass clippings — are probably in short supply compared with the needs of cropped areas, but are still useful. It might also be possible to obtain leaves from a nearby town. These can either be directly applied and worked into the soil or composted first. As with home gardeners, market gardeners can purchase manures, composts, and straw mulch, but should get volume discounts on the amounts needed for an acre or two.

---

or grass clippings used as mulch also add organic matter when they later become incorporated into the soil by plowing or by the activity of soil organisms. Some vegetable farmers use a "mow-and-blow" system where crops are grown on strips for the purpose of chopping them and spraying the residues onto an adjacent strip.

MAINTAINING SOIL BIODIVERSITY

The role of diversity is critical to maintaining a well functioning and stable agriculture. Where many different types of organisms coexist, there are fewer disease, insect, and nematode problems. There is more competition for food and more possibility that many types of predators will be found. This means that no single pest organism will be able to reach a population high enough to cause a major decrease in crop yield. We can promote a diversity of plant species growing on the land by using cover crops, intercropping, and crop rotations. However, don't forget that diversity below the soil surface is as important as diversity above ground. Growing cover crops and using crop rotations help maintain the diversity below ground, but adding manures and composts and making sure that crop residues are returned to the soil are also critical for promoting soil organism diversity.

Managing Soils and Crops to Minimize Pest Problems

Many of the practices discussed in this chapter and the other chapters in Part 2 help to reduce the severity of crop pests. It is now known that plants have very sophisticated defense mechanisms against insects and diseases. When plants are under environmental stresses caused by compact soils, droughty conditions, or excess nitrogen, they are less able to combat pests and may be even more attractive to them. On the other hand, healthy plants growing on soils with good biological diversity are able to mount a strong defense against many pests. For example, when attacked by insects they may emit chemicals that attract beneficial insects that are predators of the pest. In addition, good soil management decreases levels of pests that live in the soil.

It is well established — and known by most farmers — that crop rotation can decrease disease, insect, nematode, and weed pressures. A few other examples are given below.

✓ Insect damage can be reduced by avoiding excess nitrogen levels in soils through better nitrogen management.

✓ Root rots and severity of leaf diseases can be reduced with composts that contain low levels of available nitrogen, but still have some active organic matter.

✓ Fungal diseases of roots and insect damage are decreased by lessening soil compaction.

✓ Many pests are kept under control by competition for resources or direct antagonism (including the beneficials feeding on them). Good quantities of a variety of organic materials help maintain a diverse group of soil organisms.

✓ Root surfaces are protected from fungal and nematode attack by high rates of beneficial mycorrhizal fungi. Most cover crops help keep mycorrhizal fungi spore counts high and promote higher rates of infection by the beneficial fungi.

✓ Parasitic nematodes can be suppressed by cover crops.

✓ Residues of some cover crops, such as winter rye, reduce weed seed germination.

✓ Weed seed numbers are reduced in soils with a lot of biological activity, with both microorganisms and insects helping the process.

MANAGING SOIL PHYSICAL CONDITIONS:

Developing and maintaining an optimum physical environment. Plants thrive in a physical environment that allows roots to actively explore a large area, gets all the oxygen and water needed, and maintains a healthy mix of organisms. Although the soil's physical environment is strongly influenced by organic matter, the practices and equipment used — from tillage to planting to cultivation to harvest — have a major impact. If a soil is too wet — whether it has poor internal drainage or it receives too much water — some remedies are needed to grow high yielding and healthy crops. Also, erosion — whether by wind or water — is an environmental hazard that needs to be kept as low as possible. Erosion is most likely when the surface of a soil is bare and doesn't contain sufficient medium- to large-size water-stable aggregates. Practices for management of soil physical properties are discussed in chapters 13 to 15.

NUTRIENT MANAGEMENT

Many of the practices that build up and maintain soil organic matter also help enrich the soil with nutrients or make it easier to manage nutrients in ways that satisfy crop needs and are also environmentally sound. For example, a legume cover crop increases a soil's active organic matter and reduces erosion, but it also adds nitrogen that can be used by the next crop. Cover crops and deep-rooted rotation crops help to cycle nitrate, potassium, calcium, and magnesium that might be lost to leaching below crop roots. Importing mulches or manures onto the farm also adds nutrients along with the organic materials. However, specific nutrient management practices are needed, such as testing manure and checking its nutrient content before applying additional nutrient sources. Other examples of nutrient management practices not directly related to organic matter management include applying nutrients timed to plant needs, liming acidic soils, and interpreting soil tests to decide on the appropriate amounts of nutrients to apply (see chapters 16 to 19). Development of farm nutrient management plans and watershed partnerships also improve soil while protecting the local environment.

Many of the practices that build up and maintain soil organic matter also help enrich the soil with nutrients or make it easier to manage nutrients in ways that satisfy crop needs and are also environmentally sound.

SOURCES

Barber, S.A. 1998. Chemistry of soil-nutrient interactions and future agricultural sustainability. In *Future Prospects for Soil Chemistry* (P.M. Huang, D.L. Sparks, and S.A. Boyd, eds.). SSSA Special Publication No. 55. Soil Science Society of America. Madison, WI.

Brady, N.C., and R.R. Weil. 1999. *The Nature and Properties of Soils*. 12th ed. Macmillan Publishing Co. New York, NY.

Cavigelli, M.A., S.R. Deming, L.K. Probyn, and R.R. Harwood (eds.). 1998. *Michigan Field*

Crop Ecology: Managing Biological Processes for Productivity and Environmental Quality. Michigan State University Extension Bulletin E-2646. East Lansing, MI.

Mitchell, J., T. Hartz, S. Pettygrove, D. Munk, D. May, F. Menezes, J. Diener, and T. O'Neill. 1999. Organic matter recycling varies with crops grown. *California Agriculture* 53(4):37–40.

Oshins. C. An Introduction to Soil Health. 1999. A slide set available at the Northeast Region SARE website: www.uvm.edu/~nesare.slide. html

Topp, G.C., K.C. Wires, D.A. Angers, M.R. Carter, J.L.B. Culley, D.A. Holmstrom, B.D. Kay, G. P. Lafond, D.R. Langille, R.A. McBride, G.T. Patterson, E. Perfect, V. Rasiah, A.V. Rodd, and K.T. Webb. 1995. Changes in Soil Structure. In *The Health of Our Soils: Toward Sustainable Agriculture in Canada* (D.F. Acton and L.J. Gregorich, eds.). Center for Land and Biological Resources Research. Research Branch, Agriculture and Agri-Food Canada. Publication 1906/E. http://res.agr.ca/CANSIS/PUB-LICATIONS/HEALTH/chapter06.html

9

Animal Manures
for Increasing Organic Matter and Supplying Nutrients

The quickest way to rebuild a poor soil is to practice
dairy farming, growing forage crops, buying . . .
grain rich in protein, handling the manure properly,
and returning it to the soil promptly.

— J. L. HILLS, C. H. JONES, AND C. CUTLER, 1908

Once cheap fertilizers became widely available after World War II, many farmers, extension agents, and scientists looked down their noses at manure. People thought more about how to get rid of manure than how to put it to good use. In fact, some scientists tried to find out the absolute maximum amount of manure that could be applied to an acre without reducing crop yields. Some farmers who didn't want to spread manure actually piled it next to a stream and hoped that next spring's flood waters would wash it away. We now know that manure, like money, is better spread around than concentrated in a few places. The economic contribution of farm manures can be considerable. The value of the nutrients in manure from a 70-cow dairy farm may exceed $7,000 per year; manure from a 50-sow farrow-to-finish operation is worth about $4,000; and manure from a 20,000-bird broiler operation is worth about $3,000. The other benefits to soil organic matter build-up, such as enhanced soil structure and better diversity and activity of soil organisms, may double the value of the manure. If you're not getting the full fertility benefit from manures on your farm, you may be wasting money.

Animal manures can have very different properties, depending on the animal species, feed, bedding, and manure-storage practices. The amounts of nutrients in the manure that become available to crops also depend on what time of year the manure is applied and how quickly it is worked into the soil. In addition, the influence of manure on soil organic matter and plant growth is influenced by soil type. In other words, it's impossible to give blanket manure application recommendations. They need to be tailored for every situation.

We'll deal mainly with dairy cow manure, because there's more information about its use on cropland. We'll also offer general information about the characteristics and uses of some other animal manures.

MANURE HANDLING SYSTEMS

Solid versus Liquid

The type of barn on the farmstead frequently determines how manure is handled on a dairy farm. Dairy-cow manure containing a fair amount of bedding, usually around 13 to 20 percent dry matter, is spread as a solid. This is most common on farms where cows are kept in individual stanchions. Liquid manure-handling systems are common where animals are kept in a "free stall" barn with little bedding. Liquid manure is usually in the range of from 2 to 10 percent dry matter (90 percent or more water). Manures with characteristics between solid and liquid are usually referred to as semi-solid or slurry, depending on the method of handling.

Composting manures is becoming an increasingly popular option for farmers. By composting manure you help stabilize nutrients, have a smaller amount of material to spread, and have a more pleasant material to spread (and if neighbors have complained about manure odors, that might be a big plus). Although it's easier to compost manure that has been handled as a solid, some farmers are separating the solids from liquid manure and then irrigating with the liquid and composting the solids. For a more detailed discussion of composting, see chapter 12.

Storage of Manure

Researchers have been investigating how best to store manure to reduce the problems that come with year-round manure spreading. Storage allows the farmer to apply manure when it's best for the crop and during appropriate weather conditions. This reduces nutrient loss from the manure caused by water runoff from the field. However, significant losses of nutrients from stored manure also may occur. One study found that, during the year, dairy manure stored in uncovered piles lost 3 percent of the solids, 10 percent of the nitrogen, 3 percent of the phosphorus, and 20 percent of the potassium. Covered piles or well-contained liquid systems, which tend to form a crust on the surface, do a better job of conserving the nutrients and solids than unprotected piles. Poultry manure, with its high amount of ammonium, may lose 50 percent of its nitrogen during storage as ammonia gas volatilizes, unless precautions are taken to conserve nitrogen.

CHEMICAL CHARACTERISTICS OF MANURES

A high percentage of the nutrients in feeds passes right through animals and ends up in their manure. Over 70 percent of the nitrogen, 60 percent of the phosphorus, and 80 percent of the potassium in feeds may be available in manures for use on cropland. In addition to the nitrogen, phosphorus, and potassium contributions given in table 9.1, manures also contain significant amounts of other nutrients, such as calcium, magnesium, and sulfur. In regions where the micronutrient zinc tends to be deficient, there is rarely any deficiency on soils receiving regular manure applications.

The values given in table 9.1 must be viewed with some caution, because the characteristics of manures from even the same type of animal may vary considerably from one farm to another. Differences in feeds, mineral supplements, bedding materials, and storage systems make manure analysis quite variable. Yet, as long as feeding, bedding, and storage practices remain unchanged on a given farm, manure characteristics will be similar from year to year.

The major difference among all the manures is that poultry manure is significantly higher in nitrogen and phosphorus than the other manure types. This is partially due to the difference in feeds given poultry versus other farm animals. The relatively high percentage of dry matter in poultry manure is also partially responsible for the higher analyses of certain nutrients, when expressed on a wet ton basis.

It is possible to take the guesswork out of estimating manure characteristics; most soil-testing laboratories will now analyze manure. Manure analysis should become a routine part of the soil fertility management program on animal-based farms.

TABLE 9.1
Manure Characteristics

	DAIRY COW	BEEF COW	CHICKEN	HOG
DRY MATTER CONTENT (%)				
Solid (fresh)	13	12	25	9
Liquid (fresh, diluted)	9	8	17	6
TOTAL NUTRIENT CONTENT (APPROXIMATE)				
Nitrogen				
lbs./ton	10	14	25	10
lbs./1,000 gal.	28	39	70	28
Phosphate				
lbs./ton	5	9	25	6
lbs./1,000 gal.	14	25	70	9
Potash				
lbs./ton	10	11	12	9
lbs./1,000 gal.	28	31	33	25
Manure Equivalents				
Solid manure (tons, fresh)	20	11	5	16
Liquid manure (gal.)	7,200	4,000	1,500	5,700

—MODIFIED FROM MADISON ET AL., 1986.

EFFECTS OF MANURING ON SOILS

Effects on Organic Matter

When considering the influence of any residue or organic material on soil organic matter, the key question is the amount of solids returned to the soil. Equal amounts of different types of manures will have different effects on soil organic matter levels. Dairy and beef manure contain undigested parts of forages, as well as bedding. They, therefore, have a high amount of complex substances, such as lignin, that do not decompose readily in soils. Using this type of manure results in a much greater long-term influence on soil organic matter than does a poultry manure without bedding. More solids are commonly applied to soil with solid manure-handling systems than with liquid systems, because greater amounts of bedding are usually included.

When conventional tillage is used to grow a crop (such as corn silage), where the entire above-ground portion is harvested, research indicates that an annual application of 20 to 30 tons of the solid type of dairy manure per acre is needed to maintain soil organic matter (table 9.2). As discussed above, a nitrogen-demanding crop, such as corn, may be able to use all of the nitrogen in 20 to 30 tons of manure. If more residues are returned to the soil by just harvesting grain, lower rates of manure application will be sufficient to maintain or build up soil organic matter.

An example of how manure addition might balance annual loss is given in figure 9.1. One large Holstein "cow year" worth of manure is about 20 tons. Although 20 tons of anything is a lot, when considering dairy manure, it translates into a much smaller amount of solids. If the approximately 5,200 pounds of solid mate-

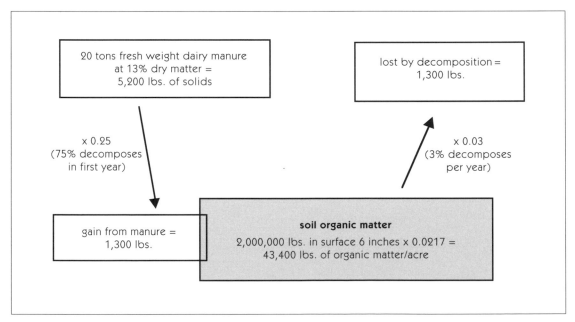

Figure 9.1 Example of dairy manure addition just balancing soil organic matter losses.

rial in the 20 tons is applied over the surface of one acre and mixed with the 2 million pounds of soil present to a 6-inch depth, it would raise the soil organic matter by about 0.3 percent. However, much of the manure will decompose during the year, so the net effect on soil organic matter will be even less. Let's assume that 75 percent of the solid matter decomposes during the first year and the carbon ends up as atmospheric CO_2. At the beginning of the following year, only 25 percent of the original 5,200 pounds, or 1,300 pounds of organic matter is added to the soil. The net effect is an increase in soil organic matter of 0.065 percent (the calcu-

Manure Influences Many Soil Properties

Application of manures causes many soil changes — biological, chemical, and physical. A few of these types of changes are indicated in table 9.2, which contains the results of a long-term experiment in Vermont with continuous corn silage on a clay soil. Manure counteracted many of the negative effects of a monoculture cropping system in which few residues are returned to the soil. Soil receiving 20 tons of dairy manure (wet weight, including bedding) maintained organic matter and CEC levels and close to the original pH (although acid-forming nitrogen fertilizers also were used). Manures, such as dairy and poultry, have liming effects and actually counteract acidification.

High rates of manure addition caused a buildup of both phosphorus and potassium to high levels. Soil in plots receiving manures were better aggregated and less dense and, therefore, had greater amounts of pore space than fields receiving no manure.

TABLE 9.2
Effects of 11 Years of Manure Additions on Soil Properties*

| | APPLICATION RATE (TONS/ACRE/YEAR) | | | |
	NONE	10 TONS	20 TONS	30 TONS
organic matter	4.3	4.8	5.2	5.5
CEC (me/100g)	15.8	17.0	17.8	18.9
pH	6.0	6.2	6.3	6.4
P (ppm)	6.0	7.0	14.0	17.0
K (ppm)	121.0	159.0	191.0	232.0
total pore space (%)	44.0	45.0	47.0	50.0

* Original levels: organic matter=5.2, pH=6.4, CEC=17.8 me/100gm, P=4 ppm, K=129 ppm. P and K levels with 20 and 30 tons of manure applied annually are much higher than crop needs (see table 19.4a).

—MAGDOFF AND AMADON, 1980; MAGDOFF AND VILLAMIL, 1977.

lation is [1,300/2,000,000] × 100). Although this does not seem like much added organic matter, if a soil had 2.17 percent organic matter and 3 percent of this was decomposed annually during cropping, then the loss would be 0.065 percent per year and the manure addition would just balance this loss.

USING MANURES

Manures, like other organic residues that decompose easily and rapidly release nutrients, are usually applied to soils in quantities judged to supply sufficient nitrogen for the crop being grown in the current year. It might be better for building and maintaining soil organic matter to apply manure at higher rates, but doing so may cause undesirable nitrate accumulation in leafy crops and excess nitrate leaching to groundwater. High nitrate levels in leafy-vegetable crops are undesirable in terms of human health, and the leaves of many plants seem more attractive to insects. In addition, salt damage to crop plants can occur from high manure application rates, especially when there is insufficient leaching by rainfall or irrigation. Very high amounts of added manures, over a period of years, also lead to high soil phosphorus levels (table 9.2). It is a waste of money and resources to add unneeded nutrients to the soil, nutrients which will only be lost by leaching or runoff, instead of contributing to crop nutrition.

Application Rates

A common per-acre rate of dairy-manure application is 10 to 30 tons fresh weight of solid, or 4,000 to 11,000 gallons of liquid manure. These rates will supply approximately 50 to 150 pounds of available nitrogen (not total) per acre. If you are growing crops that don't need that much nitrogen, such as small grains, 10 to 15 tons of solid manure should supply sufficient nitrogen per acre. For a crop that needs a lot of nitrogen, such as corn, 20 to 30 tons per acre may be necessary to supply its nitrogen needs. Low rates of about 10 tons per acre are also suggested for each of the multiple applications used on a grass hay crop. In total, grass hay crops need at least as much total nitrogen applied as does a corn crop. There has been some discussion about applying manures to legumes. This practice has been discouraged because the legume uses the nitrogen from the manure, and much less nitrogen is fixed from the atmosphere. However, the practice makes sense on animal farms where there is excess nitrogen.

For the most nitrogen benefit to crops, manures should be incorporated into the soil immediately after spreading on the surface. About half of the total nitrogen in dairy manure comes from the ammonium (NH_4^+) in urine. This ammonium represents almost all of the readily available nitrogen present in dairy manure. As materials containing urea or ammonium dry on the soil surface, the ammonium is converted to ammonia gas (NH_3) and lost to the atmosphere. If dairy manure stays on the soil surface, about 25 percent of the nitrogen is lost after one day and 45 percent is lost after four days — but that 45 percent of the total represents around 70 percent of the readily available nitrogen! This problem is significantly lessened if about ½ inch of rainfall occurs shortly after manure application, leaching ammonium from manure into the soil. Leaving manure on the soil surface is also a problem because runoff waters may carry significant amounts of nutrients from the field. When this happens, crops don't benefit as much from the manure application and surface waters become polluted. Some liquid manures — those with low solids contents — penetrate the soil more deeply. When applied at normal rates, these manures will not be as prone to lose ammonia by surface drying.

Other nutrients contained in manures, in addition to nitrogen, make important contributions to soil fertility. The availability of phosphorus and potassium in manures should be similar to that in commercial fertilizers. (However, some recommendation systems assume that only around 50 percent of the phosphorus and 90 percent of the potassium is available.) The phosphorus and potassium contributions of 20 tons of dairy manure is approximately equivalent to about 30 to 50 lbs. of phosphate and 180 to 200 lbs. of potash from fertilizers. The sulfur content as well as trace elements in manure, such as the zinc previously mentioned, also add to the fertility value of this resource.

N, P, and K in Hog, Dairy, and Beef Cattle Manures

	Feces	Urine
N	1/2	1/2
P	most	-
K	-	most

Because one-half of the nitrogen and almost all of the phosphorus is in the solids, much of these nutrients remain in sediments at the bottom when a liquid system is emptied without properly agitating the manure. On the other hand, almost all of the potassium will be applied with the liquid portion, even if it's applied without the solids. A manure system that allows signifi-

cant amounts of surface water penetration and then drainage, such as a manure stack of well-bedded dairy or beef cow manure, may lose a lot of potassium. The 20 percent leaching loss of potassium from stacked dairy manure mentioned above occurred because potassium was mostly found in the liquid portion of the manure.

Timing of Applications

Manures are best applied to annual crops, such as corn, small grains, and vegetables, in one dose just before soil tillage (unless a high amount of bedding is used, which might tie-up nitrogen for a while — see discussion of C:N in chapter 8). This allows for rapid incorporation by plow, chisel, harrow, or disk. Even with reduced tillage systems, application close to planting time is best, because the possibility of loss by runoff and erosion is reduced. It also is possible to inject liquid manures either just before the growing season starts or as a sidedress to row crops. Fall manure applications on annual row crops, such as corn, may result in considerable nitrogen loss, even if manure is incorporated. Losses of nitrogen from fall-applied manure in humid climates may be around 50 percent — resulting from leaching and denitrification before nitrogen is available to next year's crop.

Without any added nitrogen, perennial grass hay crops are constantly nitrogen deficient. Application of a moderate rate of manure — about 50 lbs. worth of available nitrogen — in early spring and following each harvest is the best way to apply manure. However, wet soils in early spring may not allow manure application without causing significant compaction.

Although the best use of manure is to apply it near the time when the crop needs the nutrients, sometimes insufficient storage capacity causes farmers to apply it at other times. In the fall, manure can be applied to grasslands that don't flood or to tilled fields that will either be fall plowed or planted to a winter cover crop. Although legal in most states, it is not a good practice to apply manures when the ground is frozen or covered with snow. The nutrient losses that can occur with runoff from winter-applied manure are both an economic loss to the farm as well as an environmental hazard. Winter spreading should be done only on an emergency basis. However, new research on frost tillage has shown that there are windows of opportunity for incorporating winter-applied manure during periods when the soil has a shallow frozen layer, 2 to 4 inches thick (see chapter 15). Farmers may use this time window to inject manure during the winter.

POTENTIAL PROBLEMS

As we all know, too much of a good thing is not necessarily good.

Excessive manure applications may cause plant-growth problems. It is especially important not to apply excess poultry manure, because the high soluble-salt content can harm plants.

Plant growth is sometimes retarded when high rates of fresh manure are applied to soil immediately before planting. This problem usually doesn't occur if the fresh manure decomposes for a few weeks in the soil and can be avoided by using a solid manure that has been stored for a year or more. Injection of liquid manure sometimes causes problems when used on poorly drained soils in wet years. The extra water applied and the extra use of oxygen by microorganisms may mean less aeration for plant roots.

When manures are applied regularly to a field to provide enough nitrogen for a crop like corn, phosphorus and potassium may build up to levels way in excess of crop needs (see table 9.2). Erosion of phosphorus-rich topsoils contributes sediments and phosphorus to streams and lakes, polluting surface waters. When very high phos-

---

Ever hear of E. Coli 0157:H7?

A bacteria strain known as E. Coli (an abbreviation that is pronounced e-COLE-eye) 0157: H7 has caused numerous outbreaks of severe illness in people who ate contaminated meat. It also has caused one known outbreak when water used to wash lettuce was contaminated with animal manure.

This particular bacteria is a resident of cows' digestive systems. It does no harm to the cow, but — probably because of the customary practice of feeding low levels of antibiotics when raising cattle — it is resistant to a number of commonly used antibiotics. This problem only reinforces the common sense approach to manure use. When using manure that has not been thoroughly composted to grow crops for direct human consumption — especially leafy crops like lettuce that grow low to the ground and root crops such as carrots and potatoes — special care should be taken. Before planting your crop, avoid problems by planning a three-month period between incorporation and harvest. For short season crops, this means that the manure should be incorporated long before planting. Although there has never been a confirmed instance of contamination of vegetables by E. Coli 0157: H7 or other disease organisms from manure incorporated into the soil as a fertility amendment, being cautious and erring on the side of safety is well justified.

---

phorus build-up occurs from the continual application of manure applied at rates to satisfy crop nitrogen needs, it may be wise to switch the application to other fields or to use strict soil-conservation practices to trap sediments before they enter a stream. Including rotation crops, such as alfalfa — that do not need manure — allows a "draw-down" of phosphorus that accumulates from manure application to grains. (However, this may mean finding another location to apply manure. For a more detailed discussion of nitrogen and phosphorus management, see chapter 17.)

Farms that purchase much of their animal feed may have too much manure to safely use on their own land. Although they don't usually realize it, they are importing large quantities of nutrients in the feed that remain on the farm as manures. If they apply all these nutrients on a small area of soil, nitrogen and phosphorus pollution of groundwater and surface water will occur. It is a good idea to make arrangements with neighbors for use of the excess manure. Another option, if local outlets are available, is to compost the manure (see chapter 14) and sell the product to vegetable farmers, garden centers, landscapers, and directly to home gardeners.

SOURCES

Elliott, L. F., and F. J. Stevenson, eds. 1977. *Soils for Management of Organic Wastes and Wastewaters.* Soil Science Society of America. Madison, WI.

Madison, F., K. Kelling, J. Peterson, T. Daniel, G. Jackson, and L. Massie. 1986. *Guidelines for Applying Manure to Pasture and Cropland in Wisconsin.* Agricultural Bulletin A3392. Madison, WI.

Magdoff, F. R., and J. F. Amadon. 1980. Yield trends and soil chemical changes resulting from N and manure application to continuous corn. *Agronomy Journal* 72:161–164. See this reference for dairy manure needed to maintain or increase organic matter and soil chemical changes under continuous cropping for silage corn.

Magdoff, F. R., J. F. Amadon, S. P. Goldberg, and G. D. Wells. 1977. *Runoff from a Low-cost Manure Storage Facility.* Transactions of the American Society of Agricultural Engineers 20:658–660, 665. This is the reference for the nutrient loss that can occur from uncovered manure stacks.

Magdoff, F. R. and R. J. Villamil, Jr. 1977. *The Potential of Champlain Valley Clay Soils for Waste Disposal.* Proceedings of the Lake Champlain Environmental Conference, Chazy, NY. July 15, 1976.

Maryland State Soil Conservation Committee. Undated. *Manure Management Handbook—A Producer's Guide.* College Park, MD.

Ontario Ministry of Agriculture and Food. 1994. *Livestock and Poultry Waste Management.* Best Management Practices Series. Available from the Ontario Federation of Agriculture, Toronto, Ontario (Canada).

Ontario Ministry of Agriculture and Food. 1997. *Nutrient Management.* Best Management Practices Series. Available from the Ontario Federation of Agriculture, Toronto, Ontario (Canada).

Soil Conservation Society of America. 1976. *Land Application of Waste Materials.* Soil Conservation Society of America. Ankeny, IA.

van Es, H.M., A.T. DeGaetano, and D.S. Wilks. 1998. Space-time upscaling of plot-based research information: frost tillage. *Nutrient Cycling in Agroecosystems* 50:85–90.

Darrell Parks
Manhattan, Kansas

Even if Darrell Parks didn't like working with pigs, he would still raise hogs on his 400-acre farm in the Flint Hills of Kansas, if only for the manure. Parks' 30 sows provide manure that makes up a key part of his soil fertility program.

Parks raises corn, milo, wheat, soybeans and alfalfa. Recently, he received organic certification, so he no longer uses any purchased fertilizer. Instead, he plants nitrogen-fixing legume cover crops such as red clover, Austrian winter peas and vetch to amend the soil and spot-treats with hog manure to help areas in need of extra fertility.

"I've been working to better utilize farm-produced manure and cover crops as well as a crop rotation and management system that will allow me to eliminate purchased fertilizer, herbicides and insecticides," says Parks, who received a grant from USDA's Sustainable Agriculture Research and Education (SARE) program to hone his use of manure on cropland.

Parks likes how manure corrects micronutrient deficiencies in his soil. He regularly tests his soils, then targets problem areas with a thicker application of manure.

Cover crops supply his nitrogen. Parks grows a legume cover crop in the winter, followed by a cash crop of milo or soybeans. On some fields, he'll grow a wheat crop planted in the fall. Before planting, he'll treat the field with manure to ensure the wheat will not lack nutrients. He follows wheat with alfalfa or clover.

At the root of Parks' program is increasing organic matter in the soil, which will improve water infiltration and soil structure. The cover crops help compensate for what Parks describes as "heavy" soils. He chooses cover crops such as sweet clover that break through compacted soil with their deep taproots. He anticipates an improvement in soil structure over the next five years as he continues to perfect his rotation.

"Back in the '20s and '30s they did some of these things and had good systems in place, then fertilizer became cheap and everyone forgot about cover crops as a possible solution," he says. "I have some fairly tight, heavy soils, and this is a way to make those soils better over time."

Parks has pushed his organic matter up above 2 percent on his sandy soils and close to or more than 3 percent on his heavy clay soils, but notes that his tillage regime makes improving organic matter content especially challenging. That's why he remains committed to his dual nutrient regime of both animal and "green" manures.

Moreover, his organic system, which should yield him more in the marketplace, demands it.

Conventional farmers in the area would benefit by emulating Parks' heavy-on-cover-crops rotation, says Ed Reznicek of the Kansas Rural Center, who works with producers to develop cropping plans. Seeding clover under wheat, or frost-seeding it, makes for good forage, increased nitrogen and much biomass, he says.

"From what I've seen, both Darrell's weed control and production seems to be increasing," Reznicek says. "He's motivated to net as much as he can from his farming operation, using a strategy of lowering costs and finding alternative markets."

10

Cover Crops

*Where no kind of manure is to be had, I think
the cultivation of lupines will be found the readiest
and best substitute. If they are sown about the middle
of September in a poor soil, and then plowed in,
they will answer as well as the best manure.*

—COLUMELLA, FIRST CENTURY, ROME

Understanding the effect of cover crops on the soil and the productivity of subsequent crops comes down to us from antiquity. Chinese manuscripts indicate that the use of green manures is probably 3,000 years old. Green manures were also commonly used in ancient Greece and Rome. There are three different terms used to describe crops grown specifically to help maintain soil fertility and productivity instead of for harvesting: green manures, cover crops, and catch crops. The terms are sometimes used interchangeably and are best thought of from the grower's perspective. A *green manure crop* is usually grown to help maintain soil organic matter and increase nitrogen availability. A *cover crop* is grown mainly to prevent soil erosion by covering the ground with living vegetation and with living roots that hold onto the soil. This, of course, is related to managing soil organic matter, because the topsoil lost during erosion contains the most organic matter of any soil layer. A *catch crop* is grown to retrieve available nutrients still in the soil following an economic crop and prevents nutrients leaching over the winter.

Sometimes it's confusing to decide which term to use — green manure, cover crop, or catch crop. We usually have more than one goal when we plant these crops during or after our main crop, and plants grown for one of these purposes may also accomplish the other two goals. The question of which term to use is not really important, so in our discussion below the term cover crop will be used.

Cover crops are usually incorporated into the soil or killed on the surface before they are mature. (This is the origin of the term green manure.) Since cover crop residues are usually low in lignin content and high in nitrogen, they decompose rapidly in the soil.

EFFECTS OF COVER CROPS

The benefits from cover crops depend on the productivity of the one that you're growing and how long it's left to grow before the soil is prepared for the next crop. The more residue you return to the soil, the better the effect on soil organic matter. The amount of residue produced by the growth of a cover crop may be very small, as little as half a ton of dry matter per acre. This adds some active organic matter, but because most decomposes rapidly after it's killed, there is no measurable effect on the total amount of organic matter present. On the other hand, good production of hairy vetch or crimson clover cover crops may yield 1½ to 2½ tons to over 4 tons per acre. If a crop like cereal rye is grown to maturity, it can produce 3 to 5 tons of residue.

A five-year experiment with clover in California showed that cover crops increased organic matter in the top 2 inches from 1.3 to 2.6 percent and in the 2- to 6-inch layer from 1 to 1.2 percent. Some researchers have found that cover crops do not seem to increase soil organic matter. Low-growing cover crops that don't produce much organic matter may not be able to counter the depleting effects of some management practices, such as intensive tillage. Even if they don't significantly increase organic matter levels, cover crops help prevent erosion and add at least some residues that are readily used by soil organisms.

Cover crops also supply nutrients to the following crop, suppress weeds, and break pest cycles. Cover crops help maintain high populations of mycorrhizal fungi spores, which helps improve inoculation of the next crop. Their pollen and nectar are important food sources for predatory mites and parasitic wasps, both important for biological control of insect pests. A cover crop also provides a good habitat for spiders, and these general insect feeders help decrease pest populations. Use of cover crops in the Southeast has reduced the incidence of thrips, bollworm, budworm, aphids, fall armyworm, beet armyworm, and white flies. Living cover crop plants and their residues also increase water infiltration into soil, thus compensating for the water that cover crops use.

SELECTION OF COVER CROPS

Before growing cover crops, you need to ask yourself some questions.

- Which type should you plant?
- When and how should you plant the crop?
- When should the crop be killed or incorporated into the soil?

When you select a cover crop, you should consider what you want to accomplish, the soil conditions, and the climate.

- Is the main purpose to add available nitrogen to the soil or to provide large amounts of organic residue?
- Is erosion control in the late fall and early spring your primary objective?
- Is the soil very acidic and infertile, with low availability of nutrients?
- Does the soil have a compaction problem? (Some species are especially good for alleviating compaction.)

- Is weed suppression your main goal?
- Which species are best for your climate? (Some species are more winter-hardy than others.)
- Will the climate and water-holding properties of your soil cause a cover crop to use so much water that it harms the following crop?

There are many types of plants that can be used as cover crops, with legumes and grasses (including cereals) the most extensively used. Leguminous crops are often very good cover crops. Summer annual legumes, usually grown only during the summer, include soybeans, peas, and beans. Winter annual legumes that are normally planted in the fall and counted on to overwinter include berseem clover, crimson clover, hairy vetch, and subterranean clover. Some, like crimson clover, can only overwinter in regions with mild frost. Hairy vetch, though, is able to withstand fairly severe winter weather. Biennials and perennials include red clover, white clover, sweet clover, and alfalfa. It should be noted that crops usually used as winter annuals are sometimes grown as summer annuals in cold, short-season regions. Also, summer annuals that are easily damaged by frost, such as cowpeas, can be grown as a winter annual in the Deep South.

One of the main reasons for selecting legumes as cover crops is their ability to fix nitrogen from the atmosphere and add it to the soil. Legumes such as hairy vetch or crimson clover that produce a substantial amount of growth may supply over 100 pounds of nitrogen per acre to the next crop. However, other legumes, such as field peas, bigflower vetch, and red clover, may supply only 30 to 80 pounds of available nitrogen.

Nonleguminous crops used as cover crops include the cereal grasses rye, wheat, oats, and barley, as well as other grass family species, such as ryegrass. Other cover crops, like buckwheat, rape, and turnips, are neither legumes nor grasses.

Some of the most important cover crops are discussed below.

LEGUMES

If you grow a legume as a cover crop, don't forget to inoculate seeds with the bacteria that live in the roots and fix nitrogen. There are various types of rhizobial bacteria that fix nitrogen. Some are specific to certain crops. There are different strains for alfalfa, clovers, soybeans, beans, peas, vetch and cowpeas. Unless you've recently grown a legume from the same general group you are currently planting, consider mixing the seeds with the appropriate commercial rhizobial inoculant before planting. The addition of sugar water to the seed-inoculant mix helps the bacteria stick to the seeds. Plant right away, so the bacteria don't dry out. Inoculums are readily available only if they are commonly used in your region. It's best to check with your seed supplier a few months before you need the inoculant, so it can be special ordered, if necessary.

Inoculum Groups

red and white clovers
crimson and berseem clovers
alfalfa, sweet clover
pea, vetch, lentils
annual medics
cowpea, lespedeza

Winter Annual Legumes

Berseem clover is an annual crop that is grown in the South during the winter. Some newer varieties have done very well in California, with "Multicut" outyielding "Bigbee." It establishes easily and rapidly and develops a dense cover, making it a good choice for weed suppression. It's also drought tolerant and re-grows rapidly when mowed or grazed. Berseem is also grown as a summer annual in the Northeast and Midwest.

Crimson clover is considered one of the best cover crops for the southeastern United States. Where adapted, it grows in the fall and winter, and matures more rapidly than most other legumes. It also contributes a relatively large amount of nitrogen to the following crop. Because it is not very winter-hardy, crimson clover is not usually a good choice for the northern portions of the South and further north. In northern regions, crimson clover can be grown as a summer annual, but that prevents an economic crop from growing during that field season. Varieties like "Chief," "Dixie," and "Kentucky Select" are somewhat winter-hardy if established early enough before winter. Crimson clover does not grow well on high pH (calcareous) or poorly drained soils.

Hairy vetch is grown in the Southeast, but is winter-hardy enough to grow well in the mid-Atlantic states and even in most of the Northeast and Midwest. Where adapted, hairy vetch produces a large amount of vegetation and fixes a significant amount of nitrogen, contributing as much as 100 pounds of nitrogen per acre or more to the next crop. Hairy vetch residues decompose rapidly and release nitrogen more quickly than most other cover crops. This can be an advantage when a rapidly growing, high-nitrogen-demand crop follows hairy vetch. Hairy vetch will do better on sandy soils than many other green manures, but needs good soil potassium levels to be most productive.

Subterranean clover is a warm climate winter annual that, in many situations, can complete its life cycle before a summer crop is planted. When used this way, it doesn't need to be suppressed or killed and does not compete with the summer crop. If left undisturbed, it will naturally re-seed itself from the pods that mature below-ground. Because it grows low to the ground and does not tolerate much shading, it is not a good choice to interplant with summer annual row crops.

Summer Annual Legumes

Cowpeas are native to central Africa and do well in hot climates. The cowpea is, however, severely damaged by even a mild frost. It is deep rooted and is able to do well under droughty conditions. It usually does better on low-fertility soils than crimson clover.

Soybeans, usually grown as an economic crop for their oil and protein-rich seeds, also can serve as a summer cover crop. They require a fertile soil for best growth. As with cowpeas, soybeans are easily damaged by frost. Soybeans, if grown to maturity and harvested for seed, do not add much in the way of lasting residues.

Biennial and Perennial Legumes

Alfalfa is a good choice for well-drained soils, near neutral in pH, and high in fertility. The good soil conditions required for the best growth of alfalfa make it a poor choice for problem situations. Where adapted, it is usually grown in a rotation for a number of years (see chapter 11). Alfalfa is commonly interseeded with small grains, such as oats, wheat, and barley, and it grows after the grain is harvested. The alfalfa variety "Nitro" can be used as an annual cover crop because it is not very winter-hardy and usually winter kills under northern conditions. Nitro continues to fix nitrogen later into the fall than winter-hardy varieties. However, it does not reliably winter kill every year, and the small amounts of extra fall growth and nitrogen fixation may not be worth the extra cost of the seed compared with perennial varieties.

Crown vetch is only adapted to well-drained soils, but can be grown under lower fertility conditions than alfalfa. It has been used successfully for roadbank stabilization and is able to provide permanent groundcover. Crown vetch has been tried as an interseeded "living mulch," with only limited success at providing nitrogen to corn. However, it is relatively easy to suppress crownvetch with herbicides to reduce its competition with corn.

Red clover is vigorous, shade tolerant, winter-hardy, and can be established relatively easily. Red clover is commonly interseeded with small grains. Because red clover starts growing slowly, the competition between it and the small grain is not usually great. Red clover also successfully interseeds with corn in the Northeast.

Sweet clover (yellow blossom) is a reasonably winter-hardy, vigorous-growing crop with an ability to get its roots into compacted subsoils. It is able to withstand high temperatures and droughty conditions better than many other cover crops. Sweet clover requires a soil pH near neutrality and a high calcium level. As long as the pH is high, sweet clover is able to grow well on low-fertility soils. It is sometimes grown for a full year or more, since it flowers and completes its life cycle in the second year. When used as a green manure crop, it is incorporated into the soil before full bloom.

White clover does not produce as much growth as many of the other legumes and is also less tolerant of droughty situations. (New Zealand types of white clover are more drought tolerant than the more commonly used Dutch white clover.) However, because it does not grow very tall and is able to tolerate shading better than many other legumes, it may be useful in orchard-floor covers or as a living mulch. It is also a common component of intensively managed pastures

GRASSES

A problem common to all the grasses is that if you grow the crop to maturity for the maximum amount of residue, you reduce the amount of available nitrogen for the next crop. This is caused by the high C:N ratio, or low percentage of nitrogen, in grasses near maturity. The problem can be avoided by killing the grass early or by adding extra nitrogen in the form of fertilizer or manure. Another way to help with this problem is to supply extra nitrogen by seeding a legume-grass mix.

Winter rye, also called cereal or grain rye, is very winter-hardy and easy to establish. Its ability to germinate quickly, together with its winter-hardiness, means that it can be planted later in the fall than most other species. Winter rye has been shown to have an allelopathic effect, which means that it can chemically suppress weeds. It grows quickly in the fall and also grows readily in the spring.

Oats are not winter-hardy. Summer or fall seedings will winter-kill under most northern conditions. This provides a naturally killed mulch the following spring and may help with weed suppression. As a mixture with one of the clovers, oats provide some quick cover in the fall. Oat stems help trap snow and conserve moisture, even after it has been killed by frost.

Annual ryegrass (not related to winter rye) grows well in the fall, if established early enough. It develops a very extensive root system and therefore provides very effective erosion control, while adding significant quantities of organic matter. It may winterkill in northern climates. Some caution is needed with annual ryegrass, because it may become a problem weed in some situations.

Sudangrass and sorghum-sudan hybrids are fast-growing summer annuals that produce a lot of growth in a short time. Because of their vigorous nature, they are good at suppressing weeds. If they are interseeded with a low-growing crop, such as strawberries or many vegetables, you may need to delay seeding so the main crop will not be severely shaded. Sundangrass is especially helpful for loosening compacted soil.

OTHER CROPS

Buckwheat is a summer annual that is easily killed by frost. It will grow better than many other cover crops on low-fertility soils. It also grows rapidly and completes its life cycle quickly. Buckwheat can grow more than 2 feet tall in the month following planting. It competes well with weeds, because it grows so fast and, therefore, is used to suppress weeds following an early spring vegetable crop. It is possible to grow more than one crop of buckwheat per year in many regions. Its seeds do not disperse widely, but it can reseed itself and become a weed. Mow or till it before seeds develop to prevent re-seeding.

Rape is a winter-hardy member of the crucifer (cabbage) family. It grows well under the moist and cool conditions of late fall, when other kinds of plants just sit there and get ready for winter. Rape is killed by harsh winter conditions in the North, but is grown as a winter crop in the middle and southern sections of the country. Members of the crucifer family do not develop mycorrhizal fungi associations, so rape will not promote mycorrhizae in the following crop.

MIXTURES OF COVER CROPS

Mixtures of cover crops offer combined benefits. The most common mixture is a grass and legume, such as winter rye and hairy vetch or oats and red clover. Mixed stands usually do a better job of suppressing weeds than a single species. Growing legumes with grasses helps compensate for the decreases in nitrogen availability for the following crop when grasses are allowed to mature. In the mid-Atlantic region, the winter rye-hairy vetch mixture has been shown to provide another advantage for managing nitrogen: When a lot of nitrate is left in the soil at the end of the season, the rye is stimulated (reducing leaching losses). When little nitrogen is available, the vetch competes better with the rye, fixing more nitrogen for the next crop.

A crop that grows erect, such as winter rye, may provide support for hairy vetch and enable it to grow better. Mowing close to the ground kills vetch supported by rye easier than vetch alone. This may allow mowing instead of herbicide use, in no-till production systems.

TIMING COVER CROP GROWTH

If you want to accumulate a lot of organic matter, it's best to grow a cover crop for the whole growing season (see figure 10.1a). This means there will be no income-generating crop grown that year. This may be useful with very infertile, possibly eroded, soils. It also may help vegetable production systems when there is no manure available and where a market for hay crops justifies a longer rotation.

Most farmers sow cover crops after the economic crop has been harvested (figure 10.1b.). In this case, as with the system shown in figure 10.1a, there is no competition between the cover crop and the main crop. The seeds can be drilled instead of broadcast, resulting in better cover crop stands. In the Deep South and in the country's mid-section, you can usually plant cover crops after harvesting the main crop. In northern areas, there may not be enough time to establish a cover crop following harvest. Even if you are able to get it established, there will be little growth in the fall to provide soil protection or nutrient uptake. The choice of a cover crop to fit between main summer crops (figure 10.1b) is severely limited in northern climates by the short growing season and severe cold. Winter rye is probably the most reliable cover crop for these conditions. In most situations, there are a range of establishment options.

The third management strategy is to interseed cover crops during the growth of the main crop (figure 10.1c). This system is especially helpful for the establishment of cover crops in short-growing-season areas. Delay seeding the cover crop until the main crop is off to a good start

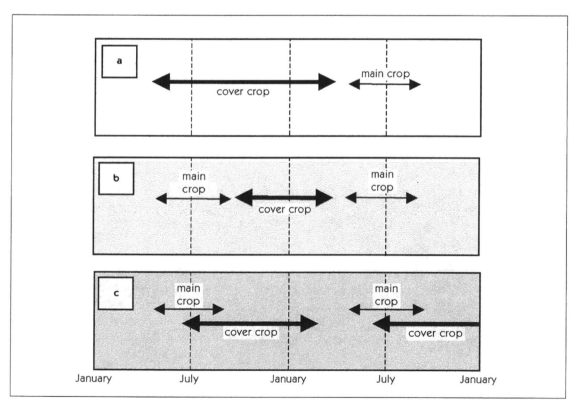

Figure 10.1 Three ways to time cover crop growth for use with a summer crop.

and will be able to grow well despite the competition. Good establishment of cover crops requires moisture and, for small-seeded crops, some covering of the seed by soil or crop residues. On the other hand, cereal rye is able to establish well without seed covering, as long as sufficient moisture is present. Farmers using this system usually broadcast seed during or just after the last cultivation. Aerial seeding, "highboy" tractors, or detasseling machines are used to broadcast green manure seed after a main crop is already fairly tall. When growing on smaller scale, seed is broadcast with the use of a hand-crank spin seeder.

When used in winter grain cropping systems, cover crops are established following grain harvest in late spring, interseeded with the grain during fall planting, or frost-seeded in early spring (figure 10.2a). With some early-maturing vegetable crops, especially in warmer regions, it is also possible to establish cover crops in late spring or early summer (figure 10.2b). Cover crops also fit into an early vegetable-winter grain rotation sequence (figure 10.2c).

No matter when you establish cover crops, they are usually killed before or during soil preparation for the next economic crop. This is done by mowing (most annuals are killed by mowing once they've flowered), plowing into the soil, with herbicides, or naturally by winter injury. Good suppression of vetch in a no-till system has been obtained with the use of a modi-

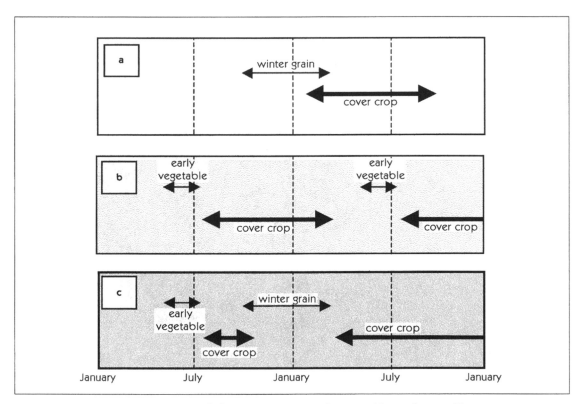

Figure 10.2 Timing cover crop growth for winter grain, early vegetable, and vegetable-grain systems.

BUILDING SOILS FOR BETTER CROPS

fied rolling stalk chopper. It is a good idea to leave a week or two between the time a cover crop is tilled in or killed and a main crop is planted. This allows some decomposition to occur and may lessen problems of nitrogen immobilization and allelopathic effects. It also may allow for the establishment of a better seedbed for small-seeded crops, such as some of the vegetables. Establishing a good seedbed for crops with small seeds may be difficult, because of the lumpiness caused by the fresh residues.

Cover crops that may become a weed problem include buckwheat, ryegrass, crown vetch, and hairy vetch.

In drier areas and on droughty soils, such as sands, late killing of a winter cover crop may result in moisture deficiency for the main summer crop. In these situations, the cover crop should be killed before too much water is removed from the soil. However, in warm climates where no-till methods are practiced, allowing the cover crop to grow longer means more residue and better water conservation for the main crop. Cover crop mulch may more than compensate for the extra water removed from the soil during the later period of green manure growth.

In very humid regions or on wet soils, the ability of an actively growing cover crop to "pump" water out of the soil by transpiration may be an advantage (see figure 14.2). Letting the cover crop grow as long as possible results in more rapid soil drying and allows for earlier planting of the main crop.

Cover crops are sometimes allowed to flower to provide bees or other beneficial insects with pollen. However, if the plants actually set seed, the cover crop may re-seed unintentionally. Cover crops that may become a weed problem include buckwheat, ryegrass, crown vetch, and hairy vetch.

INTERCROPS

Growing a cover crop between the rows of a main crop has been practiced for a long time. It has been called a living mulch, an intercrop, polyculture (if more than one crop will be harvested), and an orchard-floor cover. Intercropping has many benefits. Compared with bare soil, a groundcover provides erosion control, better conditions for using equipment during harvesting, higher water-infiltration capacity, and an increase in soil organic matter. In addition, if the cover crop is a legume, a significant buildup of nitrogen may be available to crops in future years. Another benefit is the attraction of beneficial insects, such as predatory mites to flowering plants. Less insect damage has been noted under polyculture than under monoculture.

Growing other plants near the main crop also poses potential dangers. The intercrop may harbor insect pests, such as the tarnished plant bug. Most of the management decisions for using intercrops are connected with minimizing competition with the main crop. Intercrops, if they grow too tall, can compete with the main crop for light, or may physically interfere with the main crop's growth or harvest. Intercrops may compete for water and nutrients. Using intercrops is a highly questionable practice if rainfall is barely adequate for the main crop and supplemental irrigation isn't available. One way to decrease competition is to delay seeding the intercrop until the main crop is well established. This is sometimes done in commercial fruit orchards. Soil improving intercrops established by delayed planting into annual main crops are usually referred to as cover crops. Herbicides, mowing, and partial rototilling are used to suppress the cover crop and give an advantage to the main crop. Another way to lessen competition from the cover is to plant the main crop in a relatively wide cover-free strip. This provides more distance between the main crop and the intercrop rows.

Sources

Allison, F.E. 1973. *Soil Organic Matter and its Role in Crop Production*. Elsevier Scientific Publishing Co. Amsterdam, Netherlands. In his discussion of organic matter replenishment and green manures (pp. 450–451), Allison cites a number of researchers who indicate that there is little or no effect of green manures on total organic matter, even though the supply of active (rapidly-decomposing) organic matter increases.

Hargrove, W.L. (ed.). 1991. *Cover Cops for Clean Water*. Soil and Water Conservation Society. Ankeny, IA.

MacRae, R.J., and G.R. Mehuys. 1985. The effect of green manuring on the physical properties of temperate-area soils. *Advances in Soil Science* 3:71–94.

Managing Cover Crops Profitably, 2nd Edition. 1998. Sustainable Agriculture Network, Handbook Series, #3. USDA Sustainable Agriculture Research and Education Program. Beltsville, MD. An excellent source for practical information about cover crops.

Miller, P.R., W.L. Graves, W.A. Williams, and B.A. Madson. 1989. *Cover Crops for California Agriculture*. Leaflet 21471. Division of Agriculture and Natural Resources, University of California. Davis, CA. This is the reference for the experiment with clover in California.

Pieters, A.J. 1927. *Green Manuring Principles and Practices*. John Wiley & Sons. New York, NY.

Power, J.F. (ed.). 1987. *The Role of Legumes in Conservation Tillage Systems*. Soil Conservation Society of America. Ankeny, IA.

Sarrantonio, M. 1997. *Northeast Cover Crop Handbook*. Soil Health Series, Rodale Institute, Kutztown, PA.

Smith, M.S., W.W. Frye, and J.J. Varco. 1987. Legume winter cover crops. *Advances in Soil Science* 7:95–139.

Peter Kenagy
Albany, Oregon

Peter Kenagy's vegetable rotation offers an annual window of opportunity to grow cover crops, which he has used to good effect for 15 years.

Kenagy, who farms just more than 300 acres in Oregon's fertile Willamette Valley, grows sweet corn, winter wheat, grass seed and green beans. In between those cash crops, he plants a variety of cover crops to build the soil and control weeds. Late summer and fall provide a large window after green beans, for example. In the ground just 70 days, green beans come off in July or August.

It's a perfect time, Kenagy says, to plant a summer cover crop like sudangrass, which will grow up to five feet tall before winter-killing with the first frost. The thick grass mulch continues to provide a good ground cover when he plants corn into it in the spring.

"I have a huge gap between one crop and the next," says Kenagy, who experiments with strip tillage to lessen his impact on the soil. "I have to control weeds during that period, which is just one of a number of things a cover crop does."

Kenagy also uses cover crops because they capture excess nutrients, containing them from flowing into the adjacent Willamette River. In addition, the crops catch silt from almost annual flood waters. In November 1999, for example, he had 100 acres under water.

"The more cover crop vegetation you have there, the more silt you catch," he says. "I try to get a lot of growth before fall."

Kenagy tries different covers to achieve different goals. He has planted common vetch and crimson clover to fix nitrogen in the soil and triticale, ryegrass, rape and oats to cut winter erosion and take up nutrients. The covers help aerate the soil and counter the effects of compaction.

Some years, he follows beans with oats and dwarf essex rape planted in August.

"Spring oats grow fast, and I'm shooting to get as much growth as possible in the fall to have as much coverage of the ground as possible," he says. At times, he asks a neighbor to bring in a herd of sheep to graze the oats.

Kenagy's commitment to good soil goes beyond planting cover crops. He participated in an experiment with Oregon State University researchers to test a strip-till machine that would disturb just 6 inches of soil — just one-fifth of the soil surface typically plowed. (For information about strip tillage, also called zone tillage, see chapter 15.)

The machine cuts slots through vegetative residue, which Kenagy likes to think mimics a more natural system. Grassland or forests, he points out, undergo perpetual cycles of accumulating new residue and undergoing decomposition by soil fauna.

"One of the most abusive things farmers do to the soil is till it, and most do it repeatedly," he says. "Strip till does less abuse to the soil, and keeping the residue on top is a much more natural way for it to be handled."

Kenagy was written up in his local newspaper for his commitment to forestry on the farm. He grows a mix of walnut, hazelnut, elderberry and cottonwood trees in a thick 200-foot-wide buffer along the Willamette River.

Kenagy cuts some of the trees for timber, but retains a dense hedgerow and riparian area that attracts wildlife and sops up nutrients to protect the river.

"As a society, we've made much too big a footprint on the land," he told the Oregon *Statesman Journal*. "I think it's time to make it smaller."

11

Crop Rotations

...with methods of farming in which grasses form an important part of the rotation, especially those that leave a large residue of roots and culms, the decline of the productive power is much slower than when crops like wheat, cotton, or potatoes, which leave little residue on the soil, are grown continuously.

—Henry Snyder, 1896

There are very good reasons to rotate crops. Rotating crops usually means fewer problems with insects, parasitic nematodes, weeds, and diseases caused by bacteria, viruses, and fungi. Rotations are effective for controlling insects like the corn rootworm, nematodes like the soybean cyst nematode, and diseases like root rot of field peas. In addition, rotations that include legumes supply nitrogen to succeeding crops. Growing sod-type forage grasses, legumes, and grass-legume mixes as part of the rotation also increases soil organic matter. When you alternate two crops, such as corn and soybeans, you have a very simple rotation. More complex rotations require three or more crops and a five- to 10-year (or more) cycle to complete.

Rotations are an important part of any sustainable agricultural system. Yields of crops grown in rotations are frequently about 10 percent higher than when grown in monoculture. When you grow a grain or vegetable crop following a legume, the extra supply of nitrogen  certainly helps. However, yields of crops grown in rotation are often higher than in monoculture, even when both are supplied with plentiful amounts of nitrogen. In addition, following a nonlegume crop with another nonlegume also produces higher yields than a monoculture. For example, when you grow corn following grass hay, or cotton following corn, you get higher yields than when corn or cotton are grown year after year. This yield benefit from rotations is sometimes called a rotation effect. Another important benefit of rotations is that growing a variety of crops in a given year spreads out labor needs and reduces risk caused by climate or market conditions.

99

ROTATIONS INFLUENCE SOIL ORGANIC MATTER LEVELS

You might think you're doing pretty well if soil organic matter remains the same under a particular cropping system. However, if you are working soils with depleted organic matter, you need to build up levels to counter the effects of previous practices. Maintaining an inadequately low level of organic matter won't do!

The types of crops you grow, their yields, the amount of roots produced, the portion of the crop that is harvested, and how you treat crop residues will all affect soil organic matter. Soil fertility itself influences the amount of organic residues returned, because more fertile soils grow higher-yielding crops, with more residues.

The decrease in organic matter levels when row crops are planted on a virgin forest or prairie soil is very rapid for the first five to 10 years,

but eventually, a plateau or equilibrium is reached. After that, soil organic matter levels remain stable, as long as production practices aren't changed. An example of what can occur during 25 years of continuously grown corn is given in figure 11.1. Soil organic matter levels increase when the cropping system is changed from a cultivated crop to a grass or mixed grass-legume sod. However, the increase is usually much slower than the decrease that occurred under continuous tillage.

A long-term cropping experiment in Missouri compared continuous corn to continuous sod and various rotations. More than 9 inches of topsoil was lost during 60 years of continuous corn. The amount of soil lost each year from the continuous corn plots was equivalent to 21 tons per acre. After 60 years, soil under continuous corn had only 44 percent as much topsoil as that under continuous timothy sod. A

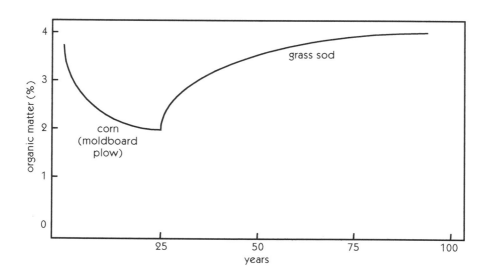

Figure 11.1 Organic matter changes in the plow layer during long-term cultivation followed by hay-crop establishment.

BUILDING SOILS FOR BETTER CROPS

six-year rotation consisting of corn, oats, wheat, clover, and two years of timothy resulted in about 70 percent as much topsoil as found in the timothy soil, a much better result than with continuous corn. Differences in erosion and organic matter decomposition resulted in soil organic matter levels of 2.2 percent for the unfertilized timothy and 1.2 percent for the continuous corn plots.

Two things happen when perennial forages (hay-type crops) are part of the rotation and remain in place for some years during a rotation. First, the rate of decomposition of soil organic matter decreases, because the soil is not continually being disturbed. (This also happens when using no-till planting, even for non-sod-type crops, such as corn.) Second, grass and legume sods develop extensive root systems, part of which will naturally die each year, adding new organic matter to the soil. Crops with extensive root systems stimulate high levels of soil biological activity. The roots of a healthy grass or legume-grass sod return more organic matter to

the soil than roots of most other crops. Older roots of grasses die, even during the growing season, and provide sources of fresh, active organic matter. Roots of plants also continually give off, or exude, a variety of chemicals that nourish nearby microorganisms.

We are not only interested in total soil organic matter — we want a wide variety of different types of organisms living in the soil. We also want to have a good amount of active organic matter and high levels of well decomposed soil organic matter, or humus, in the soil. Although most experiments have compared soil organic matter changes under different cropping systems, few experiments have looked at the effects of rotations on soil ecology. The more residues your crops leave in the field, the greater the populations of soil microorganisms. Experiments in a semiarid region in Oregon found that the total amount of microorganisms in a two-year wheat-fallow system was only about 25 percent of the amount found under pasture. Conventional moldboard plow tillage systems

TABLE 11.1
Comparison of Rotations:
Percent of Time Active Roots are Present
and Number of Species

Rotation	Years	Active Rooting Period (%)	Number of Species
Corn-soybeans	2	32	2
Dry beans-winter wheat	2	57	2
Dry beans-winter wheat/cover	2	92	3
Dry beans-winter wheat-corn	3	72	3
Corn-dry beans-winter wheat/cover	3	76	4
Sugar beets-beans-wheat/cover-corn	4	65	5

—Michigan Field Crop Ecology, 1998.

are known to decrease the populations of earthworms, as well as other soil organisms. More complex rotations increase soil biological diversity. Including perennial forages in the rotation enhances this effect.

RESIDUE AVAILABILITY

As pointed out in chapters 3, 5, and 8, more residues are left in the field after some crops than others. High residue-producing crops should be incorporated into rotations whenever possible.

SPECIES RICHNESS AND ACTIVE ROOTING PERIODS

In addition to the quantity of residues remaining following harvest, a variety of types of residues is also important. The goal should be a minimum of three different species in a rotation, with more if possible. The percent of the time that living roots are present during a rotation is also important. The period that active roots are present varies considerably, ranging from 32 percent of a corn-soybeans rotation to 57 percent of the time for a beans-wheat rotation to 76 percent of the time for a 3-year beans-wheat-corn rotation (table 11.1).

FARM LABOR AND ECONOMICS

Before discussing appropriate rotations, let's consider some of the possible effects on farm labor and finances. If you grow only one or two row crops, you must work incredibly long hours during planting and harvesting seasons. Including forage hay crops and early harvested crops, along with those that are traditionally harvested in the fall, allows farmers to spread their labor over the growing season, making the farm more eas-

ily managed by family labor alone. In addition, when you grow a more diversified group of crops, you are less affected by price fluctuations of one or two crops. This may provide more year-to-year financial stability.

Although, as pointed out above, there are many possible benefits of rotations, there are also some costs or complicating factors. It is critically important to carefully consider the farm family's labor and management capacity when exploring diversification opportunities. You may need more equipment to grow a number of different crops. There may be conflicts between labor needs for different crops; cultivation and sidedressing nitrogen fertilizer for corn in some locations might occur at the same time as harvesting hay. In addition, the more diversified the farm, the less chance for time to relax.

GENERAL PRINCIPLES

Try to consider the following principles when you're thinking about a new rotation:

1. Follow a legume forage crop, such as clover or alfalfa, with a high nitrogen-demanding crop, such as corn, to take advantage of the nitrogen supply.
2. Grow less nitrogen-demanding crops, such as oats, barley, or wheat, in the second or third year after a legume sod.
3. Grow the same annual crop for only one year, if possible, to decrease the likelihood of insects, diseases, and nematodes becoming a problem.
4. Don't follow one crop with another closely related species, since insect, disease, and nematode problems are frequently shared by members of closely related crops.

BUILDING SOILS FOR BETTER CROPS

5. Use crop sequences that promote healthier crops. Some crops seem to do well following a particular crop (for example, cabbage family crops following onions, or potatoes following corn). Other crop sequences may have adverse effects, as when potatoes have more scab following peas or oats.

6. Use crop sequences that aid in controlling weeds. Small grains compete strongly against weeds and may inhibit germination of weed seeds, row crops permit mid-season cultivation, and sod crops that are mowed regularly or intensively grazed help control annual weeds.

7. Use longer periods of perennial crops, such as a forage legume, on sloping land and on highly erosive soils. Using sound conservation practices, such as no-till planting, extensive cover cropping, or strip-cropping (a practice that combines the benefits of rotations and erosion control), may lessen the need to follow this guideline.

8. Try to grow a deep-rooted crop, such as alfalfa, safflower, or sunflower, as part of the rotation. These crops scavenge the subsoil for nutrients and water, and channels left from decayed roots can promote water infiltration.

9. Grow some crops that will leave a significant amount of residue, like sorghum or corn harvested for grain, to help maintain organic matter levels.

10. When growing a wide mix of crops — as is done on many direct marketing vegetable farms — try grouping into blocks according to plant family, timing of crops (all early season crops together, for example), type of crop (root vs. fruit vs. leaf), or crops with similar cultural practices (irrigated, using plastic mulch).

ROTATION EXAMPLES

It's impossible to recommend specific rotations for a wide variety of situations. Every farm has its own unique combination of soil and climate and of human, animal, and machine resources. The economic conditions and needs are also different on each farm. You may get useful ideas by considering a number of rotations with historical or current importance.

A five- to seven- year rotation was common in the mixed livestock-crop farms of the northern Midwest and Northeast during the first half of the 20th century. An example of this rotation is the following:

Year 1. Corn
Year 2. Oats (mixed legume/grass hay seeded)
Years 3, 4, and 5. Mixed grass-legume hay
Years 7 and 8. Pasture

The most nitrogen-demanding crop, corn, followed the pasture, and grain was harvested only two of every five to seven years. A less nitrogen-demanding crop, oats, was planted in the second year as a "nurse crop" when the grass-legume hay was seeded. The grain was harvested as animal feed and oat straw was harvested to be used as cattle bedding; both eventually were returned to the soil as animal manure. This rotation maintained soil organic matter in many situations, or at least didn't cause it to decrease too much. On prairie soils, with their very high original contents of organic matter, levels still probably decreased with this rotation.

In the corn belt region of the Midwest, a change in rotations occurred as pesticides and fertilizers became readily available and animals were fed in large feedlots, instead of on crop-producing farms. Once the mixed livestock

For many years, the western corn rootworm was effectively controlled by alternating between corn and soybeans. Recently, populations of the rootworm with a longer resting period have developed and they are able to survive the very simple rotation.

farms became grain-crop farms or crop-hog farms, there was little reason to grow sod crops. In addition, government commodity price support programs unintentionally encouraged farmers to narrow production to just two feed grains. The two-year corn-soybean rotation is better than monoculture, but it has a number of problems, including erosion, groundwater pollution with nitrate and herbicides, depletion of soil organic matter, and increased insect problems (see box). Research indicates that with high yields of corn grain there may be sufficient residues to maintain organic matter. With soybeans, residues are minimal.

The Thompson mixed crop-livestock (hogs and beef) farm in Iowa practices an alternate seven-year corn belt rotation similar to the first rotation we described. For fields that are convenient for pasturing beef cows, the Thompson rotation is as follows:

Year 1. Corn
Year 2. Soybeans
Year 3. Corn
Year 4. Oats (mixed legume/grass hay seeded)
Years 5, 6, and 7. Mixed grass-legume hay

Organic matter is maintained through a combination of practices that include the use of manures and municipal sewage sludge, green manure crops (oats and rye following soybeans and hairy vetch between corn and soybeans),

crop residues, and sod crops. These practices have resulted in a porous soil that has significantly lower erosion, higher organic matter content, and more earthworms than neighbors' fields

A four-year rotation under investigation in Virginia uses mainly no-till practices as follows:

Year 1. Corn, winter wheat no-till planted into corn stubble.
Year 2. Winter wheat grazed by cattle, foxtail millet no-till planted into wheat stubble and hayed or grazed, alfalfa no-till planted in fall.
Year 3. Alfalfa harvested and/or grazed.
Year 4. Alfalfa is harvested and/or grazed as usual until fall, then heavily stocked with animals to weaken it so that corn can be planted the next year.

This rotation follows many of the principles discussed earlier in this chapter. It was designed by researchers, extension specialists, and farmers and is very similar to the older rotation described earlier. A few differences exist — this rotation is shorter, alfalfa is used instead of clover or clover-grass mixtures, and there is a special effort to minimize pesticide use under no-till practices. Weed-control problems occurred when going from alfalfa (fourth year) back to corn. This caused the investigators to use fall tillage followed by a cover crop mixture of winter rye and hairy vetch. Some success was achieved suppressing the cover crop in the spring by just rolling over it with a disk harrow and planting corn through the surface residues with a modified no-till planter. The heavy cover crop residues on the surface provided excellent weed control for the corn.

Traditional wheat-cropping patterns for the semi-arid regions of the Great Plains and the Northwest commonly include a fallow year to allow storage of water and more mineralization of nitrogen from organic matter for use by the next wheat crop. However, the wheat-fallow system has several problems. Because no crop residues are returned during the fallow year, soil organic matter decreases unless manure or other organic materials are provided from off the field. Water infiltrating below the root zone during the fallow year moves salts through the soil to the low parts of fields. Shallow groundwater can come to the surface in these low spots and create "saline seeps," where yields will be decreased. Increased soil erosion, caused by either wind or water, commonly occurs during fallow years and organic matter decreases (at about 2 percent per year, in one experiment). In this wheat monoculture system, the build-up of grassy weed populations, such as jointed goat grass and downy brome, also indicates that crop diversification is essential.

Farmers in this region who are trying to develop more sustainable cropping systems should consider using a number of species, including deeper-rooted crops, in a more diversified rotation. This would increase the amount of residues returned to the soil, reduce tillage, and lessen or eliminate the fallow period.

A four-year wheat-corn-millet-fallow rotation under evaluation in Colorado was found to be better than the traditional wheat-fallow system. Wheat yields have been higher in this rotation than wheat grown in monoculture. The extra residues from the corn and millet also are helping to increase soil organic matter. Many producers are also including sunflower, a deep-rooting crop, in a wheat-corn-sunflower-fallow rotation. Sunflower is also being evaluated in Oregon as part of a wheat cropping sequence.

Vegetable farmers who grow a large selection of crops find it best to rotate in large blocks — with each containing crops from the same families or having similar production schedules or cultural practices. Many farmers are now using cover crops to help "grow their own nitrogen," utilize extra nitrogen that might be there at the end of the season, and add organic matter to the soil. A four- to five-year vegetable rotation might be as follows:

Year 1. Sweet corn followed by a hairy vetch/ winter rye cover crop.
Year 2. Pumpkins, winter squash, summer squash followed by a rye or oats cover crop.
Year 3. Tomatoes, potatoes, peppers followed by a vetch/rye cover crop.
Year 4. Crucifers, greens, legumes, carrots, onions, and miscellaneous vegetables followed by a rye cover crop.
Year 5. (If land is available) Oats and red clover or buckwheat followed by a vetch/rye cover crop.

Another rotation for vegetable growers uses a two- to three-year alfalfa sod as part of a six- to eight-year cycle. In this case, the crops following the alfalfa are high-nitrogen-demanding crops, such as corn or squash, followed by cabbage or tomatoes, and, in the last two years, crops needing a fine seedbed, such as lettuce, onions, or carrots. Annual weeds in this rotation are controlled by the harvesting of alfalfa a number of times each year. Perennial weed populations can be decreased by cultivation during the row-crop phase of the rotation.

Most vegetable farmers do not have enough land — or the markets — to have a multi-year hay crop on a significant portion of their land. Aggressive use of cover crops will help to maintain organic matter in this situation. Manures, composts, or other sources of organic materials, such as leaves, should also be applied every year or two to help maintain soil organic matter.

SOURCES

Anderson, S.H., C.J. Gantzer, and J.R. Brown. 1990. Soil physical properties after 100 years of continuous cultivation. *Journal of Soil and Water Conservation* 45:117–121.

Baldock, J.O., and R.B. Musgrave. 1980. Manure and mineral fertilizer effects in continuous and rotational crop sequences in central New York. *Agronomy Journal* 72:511–518.

Barber, S.A. 1979. Corn residue management and soil organic matter. *Agronomy Journal* 71:625–627.

Michigan Field Crop Ecology: Managing Biological Processes for Productivity and Environmental Quality. 1998. Cavigelli, M. A., S. R. Deming, L. K. Probyn, and R. R. Harwood (eds.). Michigan State University Extension Bulletin E-2646. East Lansing, MI.

Coleman, E. 1989. *The New Organic Grower.* Chelsea Green. Chelsea, VT. See this reference for the vegetable rotation.

Francis, C.A., and M.D. Clegg. 1990. Crop rotations in sustainable production systems. In *Sustainable Agricultural Systems* (C.A. Edwards, R. Lal, P. Madden, R.H. Miller, and G. House, eds.). Soil and Water Conservation Society. Ankeny, IA.

Gantzer, C.J., S.H. Anderson, A.L. Thompson, and J.R. Brown. 1991. Evaluation of soil loss after 100 years of soil and crop management. *Agronomy Journal* 83:74–77. This source describes the long-term cropping experiment in Missouri.

Grubinger, V.P. 1999. *Sustainable Vegetable Production: From Start-Up to Market.* Natural Resource and Agricultural Engineering Service, Ithaca, NY.

Havlin, J.L., D.E. Kissel, L.D. Maddux, M.M. Claassen, and J.H. Long. 1990. Crop rotation and tillage effects on soil organic carbon and nitrogen. *Soil Science Society of America Journal* 54:448–452.

Luna, J.M., V.G. Allen, W.L. Daniels, J.F. Fontenot, P.G. Sullivan, C.A. Lamb, N.D. Stone, D.V. Vaughan, E.S. Hagood, and D.B. Taylor. 1991. Low-input crop and livestock systems in the southeastern United States. pp. 183-205. In *Sustainable Agriculture Research and Education in the Field.* Proceedings of a conference, April 3–4, 1990. Board on Agriculture, National Research Council. Washington, D.C.: National Academy Press. This is the reference for the rotation experiment in Virginia.

National Research Council. 1989. *Alternative Agriculture.* National Academy Press. Washington, D.C. This is the reference for the rotation used on the Thompson farm.

Peterson, G.A., and D.G. Westfall. 1990. Sustainable dryland agroecosystems. In *Conservation Tillage. Proceedings of the Great Plains Conservation Tillage System Symposium*, August 21–23, 1990, Bismark, ND. Great Plains Agricultural Council Bulletin No. 131. See this reference for the wheat-corn-millet-fallow rotation under evaluation in Colorado.

Rasmussen, P.E., H.P. Collins, and R.W. Smiley. 1989. *Long-Term Management Effects on Soil Productivity and Crop Yield in Semi-Arid Regions of Eastern Oregon.* USDA-Agricultural Research Service and Oregon State University Agricultural Experiment Station, Columbia Basin Agricultural Research Center, Pendleton, OR. This describes the Oregon study of sunflowers as part of a wheat cropping sequence.

Werner, M.R., and D.L. Dindal. 1990. Effects of conversion to organic agricultural practices on soil biota. *American Journal of Alternative Agriculture* 5(1): 24–32.

Alex and Betsy Hitt
Graham, North Carolina

When the horse stable down the road went out of business, it forced Alex and Betsy Hitt to re-evaluate their farm fertility program. The Hitts, who raise 75 varieties of vegetables and an equal number of cut flowers just outside Chapel Hill, N.C., were forced to search for an alternative to horse manure to amend the soil on their five-acre farm.

The Hitts, who have made the most out of every acre, created an elaborate rotation that includes both winter and summer cover crops to supply organic matter and nitrogen, lessen erosion and crowd out weeds.

"We designed a rotation so that cover crops play a clear role," Alex Hitt says. "Many times, where other growers might say, 'I need to grow a cash crop,' we'll grow a cover crop anyway."

The Hitts stay profitable, however, thanks to a marketing plan that takes full advantage of their location near Chapel Hill, home to the University of North Carolina. Their more unusual produce such as leafy greens, leeks and rapini find a home in restaurants, and — alongside their most profitable lettuce, tomato, pepper, and flower crops — sell well at area farmers markets.

A typical year in one unit of the Hitts' rotation includes a cool-season crop, a summer cover crop such as soybeans and sudangrass, followed by a fall season cash crop and then a winter cover.

"We have made a conscious decision in our rotation design to always have cover crops," Alex Hitt says. "We have to — it's the primary source for all of our fertility. If we can, we'll have two covers on the same piece of ground in the same year."

While other farmers grow beans, corn or another profitable annual vegetable in the summer after a spring crop, the Hitts don't hesitate to take the land out of production. Instead, Alex Hitt says, their commitment to building organic matter in the soil yields important payoffs. The farm remains essentially free of soil-borne diseases, which they attribute to "so much competition and diversity" in the soil. And, despite farming on a five-percent slope, they see little or no erosion.

Alex and Betsy Hitt's Rotation
(cover crops in bold)

Year 1.	Tomatoes (half no-till)
	Oats w/ Crimson Clover
Year 2.	Cool Season Flowers
	Sudangrass w/ Soybeans
	Oats w/ Crimson Clover
Year 3.	Spring Lettuce
	Summer Flowers
	Rye w/ Hairy Vetch
Year 4.	No-till Squash
Year 5.	Over-wintered Flowers
	Sudangrass w/ Soybeans
	Rye w/ Hairy Vetch
Year 6.	Peppers (half no-till)
	Wheat w/ Crimson Clover
Year 7.	Summer Flowers
	Oats w/ Crimson Clover
Year 8.	Mixed Spring Vegetables
	Cowpeas
Year 9.	Over-wintered Flowers
	Sudangrass w/ Soybeans
	Oats w/ Crimson Clover
Year 10.	Summer Flowers
	Wheat w/ Hairy Vetch

"There are a billion benefits from cover crops," Alex Hitt says. "We have really active soil — we can see it by the good crops that we grow, and by the problems that we *don't* have."

The Hitts' rotation works well for growing flowers, a profitable direct-to-market crop that usually requires less nitrogen than vegetables. The challenge, Alex Hitt says, is in choosing the right cover prior to the next crop to get the maximum growth from the cover.

They continue to test different cover varieties — this year it's Austrian winter peas and several different clovers — in a quest for covers that are easy to establish and incorporate. The Hitts are beginning to grow some crops in a no-till system, so an easy-to-kill cover crop is paramount.

Not only does the rotation help improve soil quality, but it also goes far toward controlling weeds. The covers smother weeds by crowding and shading them out. A summer crop of cowpeas, for example, covers all bare soil. Even more effective is the Hitts' complex rotation, which confounds the weeds by varying the timing and spacing of planting and cultivation season to season.

"We either have a different crop or we're planting it differently, so we don't get the same weeds the same time every year," Alex Hitt says. "When we went to a longer rotation and changed the timing, we noticed it quickly."

The Hitts keep in touch with their soil mineral balance by testing all sections annually. They watch pH (calcium and magnesium), phosphorus and potassium levels. Keeping the proper balance in the soil, plus their complex 10-year rotation has helped reduce agricultural pests, Alex Hitt says.

"The whole system works better," he says. "We don't have many diseases and we have a lot of beneficial insects. The whole thing is really in balance, and the rotation and cover crops have a lot to do with that."

12

Making and Using Composts

*The reason of our thus treating composts of various
soils and substances, is not only to dulcify, sweeten,
and free them from the noxious qualities they otherwise
retain . . . [Before composting, they are] apter to ingender
vermin, weeds, and fungous . . . than to produce wholsome
[sic] plants, fruits and roots, fit for the table.*

—J. EVELYN, SEVENTEENTH CENTURY

Decomposition of organic materials takes place naturally in forests and fields all around us. Composting is the art and science of combining available organic wastes so that they decompose to form a uniform and stable finished product. Composts are excellent organic amendments for soils. Composting reduces bulk, stabilizes soluble nutrients, and hastens the formation of humus. Most organic materials, such as manures, crop residues, grass clippings, leaves, sawdust, and many kitchen wastes, can be composted.

The microorganisms that do much of the work of rapid composting need high temperatures, plenty of oxygen, and moisture. These heat-loving, or thermophilic, organisms work best between about 110 and 130°F. Temperatures above 140°F can develop in compost piles, helping kill off weed seeds and disease organisms, but this overheating usually slows down the process. At temperatures below 110°F, the less active mesophilic organisms take over and the rate of composting again slows down. The composting process is slowed by anything that inhibits good aeration or the maintenance of high enough temperatures and sufficient moisture.

Composting farm wastes and organic residues from off the farm has become a widespread practice. Accepting and composting lawn and garden wastes provides some income for farmers near cities and towns. They may charge for accepting the wastes and for selling compost. Some farmers, especially those without animals or sod crops, may want to utilize the compost as a source of organic matter for their own soils.

MAKING COMPOSTS

Moisture

The amount of moisture in a compost pile is important. If the materials mat and rainwater can't drain easily through the pile, it may not stay aerobic in a humid climatic zone. On the other hand, if composting is done inside a barn or under dry climatic conditions, the pile may not be moist enough to allow microorganisms to do their jobs. Moisture is lost during the active phase of composting, so it may be necessary to add water to a pile. In fact, even in a humid region, it is a good idea to moisten the pile at first, if dry materials are used. However, if something like liquid manure is used to provide a high-nitrogen material, sufficient moisture will most likely be present to start the composting process. The ideal moisture content of composting material is about 40 to 60 percent, or about as damp as a wrung-out sponge. If the pile is too dry — 35 percent or less — ammonia is lost as a gas and beneficial organisms don't repopulate the compost after the temperature moderates. Very dry, dusty composts become populated by molds instead of the beneficial organisms we want.

Types of Starting Materials

The organic materials used should have lots of carbon and nitrogen available for the microorganisms to use. High-nitrogen materials, such as chicken manure, can be mixed with high-carbon materials like hay, straw, leaves, or sawdust. Compost piles are often built by alternating layers of these materials. Turning the pile mixes the materials together. Manure mixed with sawdust or wood chips used for bedding can be composted as is. Composting occurs most easily if the average C:N ratio of the materials is about 25 to 40 parts carbon for every part nitrogen (see chapter 8 for a discussion of C:N ratios).

There are too many different types of materials that you might work with to give blanket recommendations about how much of each to mix to get the moisture content and the C:N into reasonable ranges so the process can get off to a good start. One example is given in the box on the following page.

Cornell University's web site for composting issues (http://www.cfe.cornell.edu/compost/Composting_homepage.html) features formulas to help you estimate the different proportions of the specific materials you might want to use in the compost pile. Sometimes it will work out that the pile may be too wet, too low in C:N (that means, too high in nitrogen), or too high in C:N (low in nitrogen). To balance your pile, you may need to add other materials, or change the ratios used. The examples given above can be remedied by adding dry sawdust or wood chips in the first two cases and nitrogen fertilizer in the third. If a pile is too dry, you can add water with a hose or sprinkler system.

One thing to keep in mind is that not all carbon is equally available for microorganisms. Lignin is not easily decomposed (we mentioned this when discussing soil organisms in chapter 3 and again in chapter 8, when we talked about the different effects that various residues have when applied to soil). Although some lignin is decomposed during composting — probably depending on factors such as the type of lignin and the

TABLE 12.1
Total Versus Biodegradable Carbon and Estimated C:N Ratios

MATERIAL %	CARBON	C:N%	CARBON	C:N%	LIGNIN	% CELL WALL	% NITROGEN
	(TOTAL)		(BIODEGRADABLE)				
newsprint	39	116	18	54	21	97	0.34
wheat straw	51	89	34	58	23	95	0.58
poultry manure	43	10	42	9	2	38	4.51
maple wood chips	50	51	44	45	13	32	0.97

—Richard, Trautmann & Krasny, 1996

moisture content — high amounts of carbon present as lignin may indicate that not all of carbon will be available for rapid composting. When residues contain high amounts of lignin, it means that the effective C:N can be quite a bit lower than indicated by using total carbon in the calculation (table 12.1). For some materials, there is little difference between the C:N calculated with total carbon versus using only biodegradable carbon.

It's a good idea to avoid using certain materials, such as coal ash, wood chips from pressure-treated lumber, manure from pets, and large quantities of fats, oils, and waxes. These types of materials are either difficult to compost or may result in compost containing chemicals that can harm crops.

Wood chips or bark are sometimes used as a bulking agent to provide a "skeleton" for good aeration. These materials may be recycled by shaking the finished compost out of the bulking material, which can then be used for a few more composting cycles.

Pile Size

A compost pile is a large natural convective structure — something like many chimneys all next to each other — moving oxygen into the pile as carbon dioxide, moisture, and heat rise from it. The materials need to fit together in a way that allows oxygen from the air to flow in freely. On the other hand, it is also important that not too much heat escape from the center of the pile. If small-sized particles are used, a "bulking agent" may be needed to make sure that enough air can enter the pile. Sawdust, dry leaves, hay, and wood shavings are frequently used as bulking agents. Tree branches need to be "chipped" and hay chopped so that it doesn't

Composting Animals

It is also possible to compost dead farm animals, which are sometimes a nuisance to get rid of. Chickens and even dead cows have been successfully composted. Cam Tabb, a West Virginia dairy and crop farmer, starts the process for large animals by laying the carcass that's been in the open for one day on a 3 to 4 ft bed of sawdust. Then he covers it with 3 to 4 ft of sawdust/horse manure and then turns the pile in 3 to 4 weeks. After all turning is done, he uses new base material on top.

mat and slow composting. Composting will take longer when large particles are used, especially those resistant to decay.

The pile needs to be large enough to retain much of the heat that develops during composting, but not so large and compacted that air can't easily flow in from the outside. Compost piles should be 3 to 5 feet tall and about 6 to 10 feet across the base after the ingredients have settled (see figure 12.1). (You might want it on the wide side in the winter, to help maintain the warm temperatures, while gardeners can make compost in a 3-feet tall by 3-feet wide pile in the summer.) Easily condensed material should initially be piled higher than 5 feet. It is possible to have long windrows of composting materials, as long as they are not too tall or wide.

Turning the Pile

Turning the composting residues exposes all the materials to the high-temperature conditions at the center of the pile. Although the materials at the top and on the sides of the pile are barely composting, they do provide insulation for the to quickly turn long compost windrows at large-scale composting facilities. Tractor-powered compost turners designed for composting on farms are also available.

Although turning compost frequently speeds up the process, it may also dry out the pile and cause more nitrogen loss. If the pile is too dry you might consider turning it when it's raining to help moisten it. If the pile is very wet, you might want to turn it on a sunny day. Very frequent turning may not be advantageous, because it can cause physical breakdown of important structural materials that aid natural aeration. The right amount of turning depends on a variety of factors, such as aeration, moisture, and temperature. Turn your compost pile to avoid cold, wet centers, break up clumps, and make the compost more uniform later in the process before use or marketing.

rest of the pile. Turning the pile rearranges all the materials and creates a new center. If piles are turned every time the interior reaches and stabilizes at about 140°F for a few days, it is possible to complete the composting process within months. On the other hand, if you only turn the pile occasionally, it may take a year or longer to complete. Equipment is now available

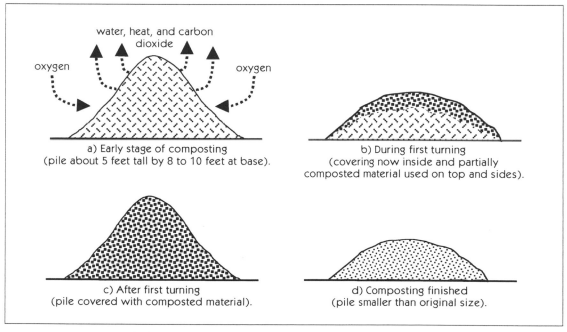

Figure 12.1 Compost pile dimensions and turning techniques.

The Curing Stage

Following high-temperature composting, the pile should be left to cure for about one to three months. Usually this is done once pile temperatures reach 105°F and high temperatures don't reoccur following turning. Curing is especially needed if the active (hot) process is short or poorly managed. There is no need to turn the pile during curing because you are not trying to stimulate maximum decomposition and there is less need for rapid oxygen entry into the pile's center when decomposition rate is slow. [However, the pile may still need turning during the curing stage if it is very large or it didn't really finish composting (determining when compost is finished is sometimes difficult), or if the pile is soaked by rain.] Curing the pile furthers aerobic decomposition of resistant chemicals and larger particles. Common beneficial soil organisms populate the pile during curing, the pH becomes closer to neutral, and ammonium is converted to nitrate. Be sure to maintain water content around 50 percent during curing to ensure that active populations of beneficial organisms develop.

It is thought that the processes that occur during the early curing process give compost some of its disease-suppressing qualities. On the other hand, beneficial organisms require sources of food to sustain them. Thus, if composts are allowed to cure for too long — depleting all the available food sources — disease suppression qualities may decrease and eventually be lost.

Disease Suppression by Composts

Research by Harry Hoitink and co-workers at Ohio State University shows that composts can suppress root and leaf diseases of plants. This suppression comes about because the plants are generally healthier (microorganisms produce plant hormones as well as chelates that make micronutrients more available) and, therefore, are better able to resist infection. Beneficial organisms compete with disease organisms for nutrients as well as directly consume the disease-causing organisms or produce antibiotics that kill bacteria. Some organisms, such as springtails and mites, "actually search out pathogen propagules in soils and devour them," according to Hoitink. In addition, he found that potting mixes containing composts "rich in biodegradable organic matter support microorganisms that induce systemic resistance in plants. These plants have elevated levels of biochemical activity relative to disease control and are better prepared to defend themselves against diseases." This includes resistance to both root and leaf diseases.

Composts rich in available nitrogen may actually stimulate certain diseases, as was found for Phytophthora root rot on soybeans, as well as Fusarium wilts and fire blight on other crops. Applying these composts many months before cropping, allowing the salts to leach away, or blending them with low nitrogen composts prior to application reduces the risk of stimulating diseases.

Composting can change certain organic materials used as surface mulches — such as bark mulches — from stimulating disease to suppressing disease.

USING COMPOSTS

Finished composts generally provide only low relative amounts of readily available nutrients. During composting, much of the nitrogen is converted into more stable organic forms, although potassium and phosphorus availability remains unchanged. However, it should be kept in mind that composts can vary significantly and some may have high levels of nitrate. Even though most composts don't supply a large amount of available nitrogen per ton, they still supply fair amounts of other nutrients in available forms and greatly help the fertility of your soil by increasing organic matter and by slowly releasing nutrients. Composts can be used on turf, in flower gardens, and for vegetable and agronomic crops. Composts can be spread and left on the surface or incorporated into the soil by plowing or rototilling. Composts also are used to grow greenhouse crops and are the basis of some potting soil mixes.

ADVANTAGES OF COMPOSTING

Composted material is less bulky than the original material, and easier and more pleasant to handle. During the composting process, carbon dioxide and water are lost to the atmosphere and the size of the pile decreases by 30 to 60 percent. In addition, many weed seeds and disease-causing organisms may be killed by the high temperatures in the pile. Unpleasant odors are eliminated. Flies, a common problem around manures and other organic wastes, are much less of a problem with composts. Composting reduces or eliminates the decline in nitrogen availability that commonly occurs when organic materials, such as sawdust or straw, are added di-

rectly to soil. Composting is also very useful for recycling kitchen wastes, leftover crop residues, weeds, and manures. Many types of local organic waste, such as apple pumice, lake weeds, leaves, and grass clippings, can be composted.

There is evidence that compost application lowers the incidence of plant root and leaf diseases (see above). In addition, the chelates and the direct hormone-like chemicals present in compost stimulate the growth of healthy plants. Then there are the positive effects on soil physical properties that are derived from improving soil organic matter. All of these factors together may help explain some of the broad benefits to plant growth that are attributed to compost.

"I don't make compost because it makes me feel good. I do it because composting is the only thing I've seen in farming that costs less, saves time, produces higher yields and saves me money."

—CAM TABB, DAIRY AND CROP FARMER, WEST VIRGINIA

If you have a large amount of organic waste but not much land, composting may be very helpful. Also, since making compost decreases the solubility of nutrients, composting may help lessen pollution in streams, lakes, and groundwater. On many poultry farms and on beef feedlots, where high animal populations on limited land may make manure application a potential environmental problem, composting may be the best method for handling the wastes. Composted

material, with about half the bulk and weight
and its higher commercial value than the manure, can be economically transported significant distances to locations where nutrients are
needed.

Without denying these good reasons to compost, there are frequently very good reasons to
just add organic materials directly to the soil,
without composting. Compared with fresh residues, composts may not stimulate as much production of the sticky gums that help hold aggregates together. Also, some uncomposted
materials have more nutrients that are readily
available to feed plants than do composts. If your
soil is very deficient in fertility, plants may need
readily available nutrients from residues. Routine use of compost as an nitrogen source may
cause high soil phosphorus levels to develop,
because of the relatively low N:P ratio. Finally,
more labor and energy usually are needed to
compost residues before applying than to simply apply the uncomposted residues directly.

SOURCES

Hoitink, H.A.J., D.Y. Han, A.G. Stone, M.S.
Krause, W. Zhang, and W.A. Dick. 1997.
Natural Suppression. *American Nurseryman.*
October 1, 1997:90–97.

Epstein, E. 1997. *The Science of Composting.*
Technomic Publishing Company.

Martin, D. L., and G. Gershuny (eds.). 1992.
*The Rodale Book of Composting: Easy Methods
for Every Gardener.* Rodale Press. Emmaus, PA.

Richard, T.L., N.M. Traubman, and M.E. Krasny.
1996. Cornell composting website. http://
www.cfe.cornell.edu/compost/

Rothenberger, R.R., and P.L. Sell. Undated. *Making and Using Compost.* University of Missouri
Extension Leaflet (File: Hort 72/76/20M). Columbia, MO.

Rynk, R. (ed.). 1992. *On Farm Composting.*
NRAES-54. Northeast Regional Agricultural
Engineering Service. Ithaca, N.Y.

Seymour, R.S. 1991. The brush turkey. *Scientific American.* Dec. 1991.

Staff of *Compost Science.* 1981. *Composting:
Theory and practice for city, industry, and farm.*
The JG Press. Emmaus, PA.

Weil, R.R., D.B. Friedman, J.B. Gruver, K.R. Is-
lam, and M.A. Stine. *Soil Quality Research at
Maryland: An Integrated Approach to Assess-
ment and Management.* Presented in Baltimore,
MD at the 1998 ASA/CSSA/SSSA meetings.
This is the source of the quote from Cam Tabb.

Cam Tabb
Kearneysville, West Virginia

Despite one of the hottest, driest summers on record, West Virginia dairy farmer Cam Tabb yielded his typical harvest of 120 bushels of corn per acre, an enviable amount to those who watched their crops wither in the fields. Neighbors wondered whether Tabb enjoyed some kind of miraculous microclimate to soar through the drought of 1999 — a severe dry spell lasting from April to September in the mid-Atlantic region — with seemingly little impact.

"I get blamed for getting more water than they got because the corn looks better," laughs Tabb, who raises 170 dairy cows and grows small grains and corn for grain and silage on 1,040 acres near Charles Town, W.V. Instead, Tabb credits his strong yields to eight years of applying composted dairy manure to his fields.

"I get a healthier plant with a better root system because my soil structure is better," he says. "So the rain that you do get really sinks in."

Tabb's compost treatments, combined with annual soil tests and rotations, have stood the test of time. Not only have his practices improved his soil and crop yields, but weather extremes like the drought of 1999 have less of an impact on his farm than those of other farmers who do not treat the soil with such care.

Tabb has come a long way since he used to pile his dairy manure on hard-packed ground and watch it ice over in the winter. After hearing a West Virginia University researcher talk about backyard composting at a town meeting, Tabb realized he needed to add a carbon source and turn the piles to encourage aeration. Once he began mixing in sawdust from horse stalls and turning the piles, he was on his way to becoming a master composter.

Now, after years of fine-tuning his operation, Tabb can talk about compost for hours. He says amending soil with compost in the spring leads to healthy plants for the rest of the season. Moreover — here he gets especially animated — composting reduces his volume of manure by half.

"The crop response and the reduction of manure volume is what keeps me doing this," he says.

Over the last few years, Tabb has worked with scientists to research the advantages of using compost in his grain fields. One scientist compared an acre of corn grown in soil amended with Tabb's compost against an acre of unamended soil on his farm. The difference in yields was extraordinary: up to 123 bushels of corn in the compost plot versus between 6 and 42 bushels in the conventional tract.

In another experiment, Tabb mixed 10,000 pounds of fish into his piles for a scientist who needed to dispose of fish that turned up with a bacterial disease during an aquaculture experiment. Contrary to what a visitor might expect, the compost turned out to be rich — and odorless. Tabb says the Native Americans who taught early Pilgrims to drop a dead fish into their soil before planting were on to something.

"I built a windrow 60 feet long and 15 feet high," he recalls. "A month later, I would never have known the fish were there."

Tabb grazes his herd on ryegrass pastures between March and July. When the cows are in the barn the remainder of the year, he collects the manure and composts it. He applies between 18 and 24 tons of compost to his crop fields, depending on soil test results, just once every three years. That way, he reduces the number of tractor trips

— and therefore, soil compaction — and still supplies all of the necessary nutrients, except extra nitrogen. Measurements show his compost supplies about 9 pounds of nitrogen, 12 pounds of phosphorus and 15 pounds of potash per ton.

"I use all I produce," says Tabb, whose many piles of compost stretch across a portion of his farm. The nutrient-supplying power of compost more than compensates for the time Tabb spends making his black gold. He uses a mix of dairy manure, leaves, grass, wood shavings, sawdust and horse manure, then turns the pile three to four times. He monitors pile temperatures and turns when it gets up to about 150 degrees, but doesn't follow any strict rules.

"I do it the lazy way," he says. "I turn when I get good and ready. It's not an extra hard operation." Turning, which Tabb does with a front-end loader, pays for itself by reducing the pile.

Those extreme temperatures in the piles, which reach as high as 150 degrees, kill both pathogens and weed seeds. In the spring, Tabb spreads a thin layer of compost on part of his fields and plants through it. When the soil is compacted, Tabb works the compost into the ground to help loosen it.

Tabb sees other benefits, too. He is especially proud of his up to 7-percent organic matter, pushed to such levels from eight years of using compost. The soil takes on a more spongy feel; Tabb says he sees little to no runoff from his compost-treated fields. It also attracts worms. In a study comparing two, one-acre blocks, researchers found 23 worms per square foot in compost-treated soil versus less than one worm per square foot in untreated soil on Tabb's farm.

While Tabb reduces pesticide use thanks to compost and a small grain-beans-corn-cover crop rotation, he continues to spot-treat with a herbicide to control weeds. He plants using no-till strategies, often through cover crop residue and compost.

Not only has Tabb changed his mind about compost, but his neighbors are starting to take notice. "People really thought I had lost my mind on this project," he says. Recently, however, a neighbor asked him for compost for his wife's garden.

"Before I handled the manure as a waste, not a resource," he says. "Now, everything just grows better."

13

Reducing Soil Erosion

So long! It's been good to know you.
This dusty old dust is a gettin' my home.
And I've got to be drifting along.

—Woody Guthrie, 1940

The dust storms that hit the center of the U.S. during the 1930s were responsible for one of the great migrations in our history. As Woody Guthrie pointed out in his songs, soil erosion was so bad that people saw little alternative to abandoning their farms. They moved to other parts of the country in search of work. Although changed climatic conditions and agricultural practices improved the situation for a time, there was another period of accelerated wind and water erosion during the 1970s and 1980s.

Erosion by wind and water has occurred since the beginning of time. Although we should expect some soil loss to occur on almost all soils, agriculture often increases erosion. Erosion is the major hazard or limitation to the use of about one-half of all cropland in the United States! On much of this land, erosion is occurring fast enough to reduce future productivity. As we discussed earlier, erosion is also an organic matter issue, because it removes the soil layer highest in organic matter, the topsoil. The soil removed from fields also has huge negative effects off the farm, as sediment accumulates in streams, rivers, and reservoirs or blowing dust reaches towns and cities.

A small amount of erosion is acceptable, as long as new topsoil can be created as rapidly as soil is lost. The maximum amount of soil that can be lost to erosion each year, while still maintaining reasonable productivity is called the soil loss tolerance or T value. For a deep soil with a rooting depth of greater than 5 feet, the T value is 5 tons per acre each year. Although this sounds like a huge amount of soil loss, keep in mind that the weight of an acre of soil to 6 inches depth is about 2 million pounds, or 1,000 tons. So 5 tons is equivalent to about .03 inches ([5/1,000] × 6 inches = 0.03 inch), enough to fill 200 bushel baskets with soil. If soil loss continues at this rate, at the end of 33 years about 1 inch will be lost. On deep soils with good management of organic matter, the rate of topsoil creation can balance this loss. The soil loss tolerance amount is gradually reduced for soils with less rooting depth. When the rooting depth is shallower than 10 inches, the T value is about 1 ton per acre each year. This is the same as 0.006 inch per year and is equivalent to 1 inch of loss in 167 years.

When your soil loss is greater than the tolerance value, productivity suffers in the long run. Yearly losses of 10 or 15 tons or more per acre occur in many fields. Management practices are available to help reduce runoff and soil losses. For example, researchers in Washington state found that erosion on winter wheat fields was about 4 tons each year when a sod crop was included in the rotation, compared to about 15 tons when sod was not included. An Ohio experiment where runoff from conventionally tilled and no-till continuous-corn fields was monitored showed that over a four-year period, runoff averaged about 7 inches of water each year for conventional tillage and less than one-tenth (0.1) inch for the no-till planting system.

SOLVING EROSION PROBLEMS

Effective erosion control is possible without compromising crop productivity. However, controlling erosion is not always easy. It may require considerable investment (as with terracing) or new management strategies (as with no-till systems). The numerous approaches to controlling erosion can be generally grouped into structural

Erosion: A Short-Term Memory Problem?

It's difficult to fully appreciate erosion's damage potential, because the most severe erosion occurs during rare weather events and climate anomalies. Wind erosion during the dust-bowl days of the 1930s was especially damaging, resulting from several extremely dry years in a row. And about one-third of water erosion damage that occurs in a particular field during a 30-year period commonly results from a single extreme rainfall event! We must do our best to adequately protect our soils from the damage that weather extremes can cause.

solutions and agronomic management practices. Structures for reducing erosion generally involve engineering practices, where an initial investment is made to build terraces, diversion ditches, drop structures, etc. Agronomic practices to reduce erosion focus on changes in soil and crop management. Appropriate conservation methods may vary among fields and farms. Recently, there has been a clear trend away from structural measures in favor of agronomic management practices. The primary reasons for this change are:

- Management measures help control erosion, while also improving soil quality and crop productivity.
- Significant advances have been made in farm machinery and methodologies for alternative soil and crop management.
- Structures generally focus on containing runoff and sediment once erosion has been initiated, whereas management measures try to prevent erosion from occurring in the first place.

- Structures are often expensive to build and maintain.
- Most structures do not reduce tillage erosion.

For long-term sustainability of crop production, use of agronomic management practices is usually preferred, although structural measures can effectively complement them.

Erosion reduction works by either decreasing the shear forces of water and wind or by keeping soil in a condition that can't erode easily. Many conservation practices actually provide both. The soil organic matter management practices we discussed in the earlier chapters all reduce erosion. We'll also briefly cover other important practices for keeping erosion to a minimum.

Reduced Tillage

Transition to tillage systems that increase surface cover (chapter 15) is probably the single most effective and economic approach to reducing erosion. Restricted and no-till regimes succeed in many cropping systems by providing similar or even better economic returns than conventional tillage, while providing excellent erosion control. Maintenance of residues on the soil surface and the lack of soil loosening by tillage greatly reduce dispersion of surface aggregates by raindrops and runoff waters. The effects of wind on surface soil are also greatly reduced by leaving crop stubble on untilled soil and anchoring the soil with roots. These measures facilitate infiltration of precipitation where it falls, thereby reducing runoff and increasing plant water availability.

In cases where tillage is necessary, reducing its intensity and leaving some residue on the surface slows down the loss of soil organic matter and aggregation. Less tillage promotes higher infiltration rates and reduces runoff and erosion. Leaving a rougher soil surface by eliminating secondary tillage passes and packers that crush natural soil aggregates may significantly reduce runoff and erosion losses by preventing surface sealing after intense rain.

Reducing or eliminating tillage also diminishes tillage erosion and keeps soil from being moved downhill. The gradual losses of soil from upslope areas exposes denser subsoil and may in many cases further aggravate runoff and erosion. It is, however, possible to gradually reverse the effects of tillage erosion caused by using a moldboard plow. Because the plow moves soil forward and to the side, topsoil can be gradually moved back up the slope, if plowing is performed diagonally to the slope in the uphill direction, with the soil being thrown 45 degrees to the front/right of its original location. Of course, this approach may not give good soil inversion during plowing and does not address water and wind erosion concerns.

Significance of Soil Residues

Reduced-tillage and no-tillage practices result in less soil disturbance and leave significant quantities of residue on the surface. Surface residues are important because they intercept raindrops and can also slow down water running over the surface. The amount of residue on the surface may be close to zero for the moldboard plow while continuous no-till planting may leave 90 percent or more of the surface covered. Other reduced-tillage systems, such as chiseling and disking (as a primary tillage operation), typically leave more than 30 percent of the surface covered.

Adding Organic Materials

Maintaining good soil organic matter levels helps keep topsoil in place. A soil with more organic matter usually has better tilth and less surface crusting. This means that more water is able to infiltrate into the soil instead of running off the field, taking soil with it. When you build up organic matter, you help control erosion by making it easier for rainfall to enter the soil.

Adding organic materials regularly to soils also results in larger and more stable soil aggregates. Larger aggregates are not eroded by wind or water as easily as smaller ones. Surface residue mulches provide both physical protection of the soil surface from raindrop impact, as well as food for large numbers of earthworms.

The adoption rate for no-till practices is lower for livestock-based farms than for grain and fiber farms. Manures may need to be incorporated into the soil for best use of nitrogen, protection from runoff, and odor control. Also, the severe compaction sometimes resulting from use of heavy liquid manure spreaders on very moist soils may need to be lessened by tillage. Direct injection of liquid organic materials in a zone or no-till system is generally an option but requires additional equipment investments.

Cover Crops

Cover crops decrease erosion and increase water infiltration in a number of ways. Cover crops add organic residues to the soil and help maintain tilth and organic matter levels. Cover crops frequently grow during seasons when the soil is especially susceptible to erosion, such as the early spring. Their roots help to bind soil and hold it in place. Because raindrops lose most of their energy when they hit leaves and drip to the ground, less soil crusting occurs. Cover crops are especially effective in reducing erosion if they are cut and mulched, rather than incorporated. See chapter 10 for more information about cover crops.

Perennial Rotation Crops

Grass and legume forage crops can help lessen erosion because they maintain a cover on most of the soil surface for the whole year. Their extensive root systems hold soil in place. Ideally, such rotations are combined with reduced and no-tillage practices for the annual crops. Permanent sod is a very good choice for steep soils or other soils that erode easily.

Other Practices and Structures For Soil Conservation

Diversion ditches are frequently helpful for channeling water away from the field without flowing over the entire area. Grassed waterways for diversion ditches and other field water channels do not reduce erosion from all of the field, but they do keep sediments on the field and reduce scouring of the channels. Grassed waterways help prevent surface water pollution by filtering sediments out of runoff before they reach a stream or pond.

Tilling and planting along the contour is a simple practice that helps control erosion. When you work along the contour, instead of up and down slope, wheel tracks and depressions caused by the plow, harrow, or planter will retain runoff water in small puddles and allow it to slowly infiltrate. This approach is not very effective when dealing with steep erodible lands and also does not reduce tillage erosion.

Alternating strips of row crops and perennial forages along the contour, referred to as strip cropping, is an effective way of reducing erosion losses. In this system, erosion from the row crop is not allowed to worsen over long, unprotected slopes because the sediments are filtered out of runoff when the water reaches the sod of the forage crop. This conservation system is generally effective in fields with moderate erosion potential, and on farms with use for both row and sod crops (for example, dairy farms). Research indicates that crop yields may be slightly higher when crops are grown in strips, rather than in the entire field. The increase in yield is probably due to better use of light and soil where the different strips meet.

Terracing soil in hilly regions is an expensive practice, but one which results in a more gradual slope and greatly reduced erosion. Well-constructed and maintained structures can last a long time, frequently making the high initial investment worthwhile.

Wind erosion is reduced by most of the same practices that reduce water erosion — reduced tillage or no-till, cover cropping, and perennial rotation crops. In addition, practices that increase roughness of the soil surface diminish the effects of wind erosion. The resulting increase in turbulent air movement near the land surface reduces the wind's shear and its ability to sweep up soil material. Therefore, fields subjected to wind erosion may be rough-tilled. Also, tree shelterbelts planted at regular distances perpendicular to the main wind direction act as windbreaks and are very effective in reducing wind erosion losses.

SOURCES

American Society of Agricultural Engineers. 1985. *Erosion and Soil Productivity.* Proceedings of the national symposium on erosion and soil productivity, December 10–11, 1984, New Orleans, Louisiana. American Society of Agricultural Engineers Publication 8-85. St. Joseph, MI.

Edwards, W.M. 1992. Soil structure: Processes and management. pp. 7–14. In *Soil Management for Sustainability* (R. Lal and F.J. Pierce, eds.). Soil and Water Conservation Society. Ankeny, IA. This is the reference for the Ohio experiment on the monitoring of runoff.

Lal, R., and F.J. Pierce (eds.). 1991. *Soil Management for Sustainability.* Soil and Water Conservation Society. Ankeny, IA.

Ontario Ministry of Agriculture, Food, and Rural Affairs. 1997. *Soil Management.* Best Management Practices Series. Available from the Ontario Federation of Agriculture, Toronto, Ontario (Canada).

Reganold, J.P., L.F. Elliott, and Y.L. Unger. 1987. Long-term effects of organic and conventional farming on soil erosion. *Nature* 330:370–372. This is the reference for the Washington State study of erosion.

Smith, P.R. and M.A. Smith. 1998. Strip intercropping corn and alfalfa. *Journal of Production Agriculture* 10:345–353.

Soil and Water Conservation Society. 1991. *Crop Residue Management for Conservation.* Proceedings of a national conference, August 8–9, 1991, Lexington, KY. Soil and Water Conservation Society. Ankeny, IA.

United States Department of Agriculture. 1989. *The Second RCA Appraisal: Soil Water, and Related Resources on Nonfederal Land in the United States, Analysis of Conditions and Trends.* Government Printing Office. Washington, DC.

14

Preventing and Lessening Compaction

*A lasting injury is done by ploughing
land too wet.*

—S. L. DANA, 1842

We've discussed the benefits of cover crops, rotations, reduced tillage, and organic matter additions for improving soil structure. However, these practices still may not prevent compacted soils unless specific steps are taken to reduce the impact of heavy loads from field equipment and inappropriately timed field operations. The causes of compaction were discussed in chapter 6, and in this chapter we'll discuss strategies to prevent and lessen soil compaction. The first step is to decide whether compaction is a problem and which type is affecting your soils. The symptoms, as well as remedies and preventive measures, are summarized in table 14.1.

CRUSTING AND SURFACE SEALING

Crusting and surface sealing are easy to see at the soil surface after heavy rains in the early growing season, especially with clean-tilled soil. Keep in mind that it may not happen every year, if heavy rains do not occur before the plant canopy protects the soil from direct raindrop impact. Certain soil types, such as sandy loams, are particularly susceptible to crusting. Their aggregates usually aren't very stable and, once broken down, the small particles fill in gaps between the larger particles.

The impact of surface crusting is most damaging when heavy rains occur between planting and seedling emergence. The hard surface that forms may delay seedling emergence and growth until the crust mellows with the next rains. If such showers do not occur, the crop may be set back considerably. Crusting and sealing of the soil surface also reduce water infiltration capacity. This increases runoff and erosion, and lessens the amount of available water for crops.

Reducing Surface Crusting

Crusting is a symptom of poor soil structure that develops especially with intensively and clean-tilled soils. As a short-term solution, farmers sometimes use tools, such as rotary hoes, to break up the crust. The best long-term approach is to reduce tillage intensity, use tillage systems that leave residue or mulch on the surface, and improve aggregate stability with organic matter additions. Even residue covers as low as 30 percent will greatly reduce crusting and provide important pathways for water entry. A good heavy-duty conservation planter — with rugged coulter blades for in-row soil loosening, tine wheels to remove surface residue from the row, and accurate seed placement — is a key implement because it can successfully establish crops without intensive tillage (see chapter 15). Reducing tillage and maintaining significant amounts of surface residues not only prevent crusting, but also rebuild the soil by reducing decomposition of organic matter. Practices that improve soil structure, such as cover cropping, rotations with perennial crops, and adding organic materials, also help reduce crusting problems. Soils with very low aggregate stability may sometimes benefit from surface applications of gypsum (calcium sulfate). The added calcium and the effect of the greater salt concentration in the soil water both promote aggregation.

PLOW LAYER AND SUBSOIL COMPACTION

Deep wheel tracks, extended periods of saturation, or even standing water following a rain or irrigation may indicate plow layer compaction. Compacted plow layers also tend to be extremely cloddy when tilled. A shovel can be used to vi-sually evaluate soil structure and rooting. This is best done when the crop is in an early stage of development, but after the rooting system had a chance to get established. Digging can be very educational and provide good clues to the quality of the soil. If you find a dense rooting system with many fine roots that protrude well into the subsoil, you probably do not have a compaction problem. Compacted soil shows little aggregation, is more difficult to dig, and will dig up in large clumps rather than granules. Compare the difference between soil and roots in wheel tracks and nearby areas.

Roots in a compacted plow layer are usually stubby and have few root hairs. They also often follow crooked paths as they try to find zones of weakness. Rooting density below the plow layer is an indicator for subsoil compaction. Roots are almost completely absent from the subsoil below severe plow pans and often move horizontally above the pan (figure 6.6). Keep in mind, however, that shallow rooted crops, such as spinach and some grasses, may not necessarily experience subsoil compaction problems under those conditions.

Compaction also may be recognized by observing crop growth. A poorly structured plow layer will settle into a dense mass after heavy rains, leaving few large pores for air exchange. If soil wetness persists, anaerobic conditions may occur, causing reduced growth and denitrification (exhibited by leaf yellowing), especially in areas that are imperfectly drained. In addition, these soils may "hardset" if heavy rains are followed by a drying period. Crops in their early growth are very susceptible to these problems (because roots are still shallow) and commonly go through a noticeable period of stunted growth on compacted soils.

TABLE 14.1
Types of Compaction and Their Remedies

COMPACTION TYPE	INDICATIONS	REMEDIES/PREVENTION
Surface crusting	Breakdown of surface aggregates and sealing of surface. Poor seedling emergence. Accelerated runoff and erosion.	Reduce tillage intensity. Leave residues on surface. Add other sources of organic matter (manures, composts). Grow cover crops.
Plow layer	Deep wheel tracks. Prolonged saturation or standing water. Poor root growth. Hard to dig and resistant to penetrometer. Cloddy after tillage.	Plow with moldboard or chisel plow, but reduce secondary tillage. Do primary tillage before winter (if no erosion danger exists). Use zone builders. Increase organic matter additions. Use cover crops or rotation crops that can break up compact soils. Use better load distribution. Use controlled traffic. Don't travel on soils that are wet. Improve soil drainage.
Subsoil	Roots can't penetrate subsoil. Resistant to penetrometer.	Don't travel on soils that are wet. Improve soil drainage. Deep tillage with a subsoiler. Use cover crops or rotation crops that penetrate compact subsoils. Use better load distibution. Use controlled traffic. No wheels in open furrows.

Reduced growth because of compaction affects the crop's ability to fight or compete with pathogens, insects or weeds. These pest problems may, therefore, become more apparent simply because the crop is weakened. For example, compacted soils that are put into a no-till system may initially show greater weed pressure because the crop is unable to effectively compete. Also, dense soils that are poorly aerated are more susceptible to infestations of certain soil-borne pathogens, such as *Phytophthora* during wet periods.

Preventing or Lessening
Soil Compaction

Preventing or lessening soil compaction generally requires a comprehensive, long-term approach to addressing soil health issues and rarely gives immediate results. Compaction on any particular field may have multiple causes and the solutions are often dependent on the soil type, climate and cropping system. Let's go over some general principles of how to solve these problems.

Tillage is a problem, but sometimes can be a solution. Tillage can either cause or lessen problems with soil compaction. Repeated intensive tillage reduces soil aggregation and compacts the soil over the long term, causes erosion and loss of topsoil, and may bring about the formation of plow pans. On the other hand, tillage can relieve compaction by loosening the soil and creating pathways for air and water movement and root growth. This relief, as effective as it may be, is temporary. Tillage may need to be repeated in the next growing seasons if soil management and traffic patterns stay the same.

Over time, farmers frequently use more intense tillage to offset the problems of cloddiness associated with compaction of the plow layer. The solution to this problem is not necessarily to stop tillage altogether. Compacted soils frequently become "addicted" to tillage and going "cold turkey" to a no-till system with a seriously degraded soil may result in failure. Practices that perform some soil loosening with minor disturbance at the soil surface help in the transition from a tilled to an untilled system. This may include a zone-building tool (figure 14.1a) with narrow shanks that disturb soil only where future plant rows will go. Also, paraplows (figure 14.1b) that loosen the soil by lifting it from underneath can relieve some compaction. Another

Lessening and preventing soil compaction is important to improving soil health. The specific approaches:

✓ should be selected based on where the compaction problem occurs (subsoil, plow layer, or surface);
✓ must fit the soil and cropping system and their physical and economic realities; and
✓ are influenced by other management choices, such as tillage system and use of organic matter amendments.

approach may be to gradually reduce tillage intensity through the use of tillage tools that leave residue on the surface (for example, chisels with straight points, or specifically designed for high-residue conditions) and a good planter that ensures good seed placement even with minimal secondary tillage. Such a system reduces organic matter losses and erosion over the long term and — through better germination rates — may produce more crop residues.

Deep tillage (subsoiling) is a method to alleviate compaction below the depth of normal tillage, although it is often erroneously seen as a cure for all types of soil compaction. It is a rather costly and energy-consuming practice that should not be done regularly. (Practices such as "zone building" and paraplowing also may loosen the soil below the plow layer, but are less rigorous and leave residue on the surface.) Deep tillage may be beneficial on soils that have developed a plow pan. Simply shattering this pan allows for deeper root exploration. To be effective, deep tillage needs to be performed when the entire depth of tillage is sufficiently dry and in the friable state. The practice tends to be more effective on coarse-textured soils (sands, gravels),

a) zone-builder

b) paraplow

Figure 14.1 Tillage implements that loosen soil with minimum surface disturbance.

as crops on those soils respond better to deeper rooting. The entire subsoil of fine-textured soils is often hard, so the effects of deep tillage are then less beneficial and in some cases even harmful. After performing deep tillage, it is important to prevent future recompaction of the soil by heavy loads and plows.

Better attention to working and traveling on the soil. Compaction of the plow layer or subsoil is often the result of working or traveling on a field when it is too wet. The first step when addressing compaction is to evaluate all traffic and practices that occur on a field during the year and determine which field operations are likely to be most damaging. The main criteria should be:

- The soil moisture condition under which the traffic occurs; and
- The relative compaction effects of various types of field traffic (mainly defined by equipment weight and load distribution).

For example, with a late-planted crop, soil conditions during tillage and planting may be gener-

ally dry, and minimal compaction damage occurs. Likewise, mid-season cultivations usually do little damage, because conditions are usually dry and the equipment tends to be light. However, if the crop is harvested under wet conditions, heavy harvesting equipment and uncontrolled traffic by trucks that transport the crop off the field will do considerable compaction damage. In this scenario, emphasis should be placed on improving the harvesting operation. In another scenario, a high-plasticity clay loam soil is often spring-plowed when still too wet. Much of the compaction damage may occur at that time and alternative approaches to tillage should be a priority.

Better load distribution. Improving the design of field equipment may help reduce compaction problems by better distributing vehicle loads. The ultimate example of this is the use of tracks, like those on a bulldozer, which especially reduce the potential for subsoil compaction. (Beware! Tracked vehicles may tempt a farmer to traffic the land when it's still too wet. Tracked vehicles have better flotation and traction, but still cause compaction damage, especially through smearing under the tracks.) Plow layer compaction also can be reduced by lowering the inflation pressure of tires. A rule of thumb: cutting tire inflation pressure in half doubles the size of the tire footprint and cuts the contact pressure on the soil in half.

Use of multiple axles reduces the load on tires. Even though the soil receives more tire passes, the resulting compaction is significantly reduced. Using large, wide tires with low inflation pressure also helps decrease the compaction effect of loads on soil. Use of dual wheels similarly reduces compaction by increasing the footprint, although this is less effective for reducing subsoil compaction, because the pressure cones from neighboring tires (figure 6.11) merge at shallower depths. Dual wheels are very effective at increasing traction, but again, pose a danger because of the temptation (and ability) to do field work under relatively wet conditions. Duals are not recommended on tractors performing seeding/planting operations because of the larger footprint.

In colder climates, the tilth of compacted plow layers can be improved by rough-tilling the soil in the fall. The freeze-thaw action on the loosened soil helps create new aggregates.

Improved soil drainage. Fields that are imperfectly drained often have more severe compaction problems, because wet conditions persist and it is almost impossible to prevent traffic or tillage under those conditions. Improving drainage may go a long way toward preventing and reducing compaction problems on poorly drained soils. Subsurface (tile) drainage improves timeliness of field operations, helps dry the subsoil and, thereby, reduces compaction in deeper layers.

Clay soils often pose the greatest challenge with respect to compaction, because they remain in the plastic state for extended periods in the spring. After the top inch near the soil surface dries out, it becomes a barrier that greatly reduces further evaporation losses (figure 14.2). This keeps the soil below in a plastic state, preventing it from being worked or trafficked without causing excessive smearing and compaction damage. For this reason, farmers often fall-till clay soils. A better approach might be to use cover crops to dry the soil in the spring. When a crop like winter rye grows rapidly in the spring, its roots effectively pump water from layers below the soil surface and allow the soil to transition from the plastic to the friable state (figure 14.2). Because these soils have high moisture-holding capacity, there is normally little concern about cover crops depleting water for the following crop.

Cover and rotation crops. Cover and rotation crops can significantly reduce soil compaction. The choice of cover/rotation crop should be defined by the climate, cropping system, nutrient needs, and the type of soil compaction. Perennial crops commonly have active root growth early in the growing season and can reach into the compacted layers when they are still wet and relatively soft. Grasses generally have shallow, dense fibrous root systems that have a very beneficial effect on the surface layer, but don't help much with subsoil compaction. Crops with deep taproots, such as alfalfa, have fewer roots at the surface, but can penetrate into a compacted subsoil. In many cases, a combi-

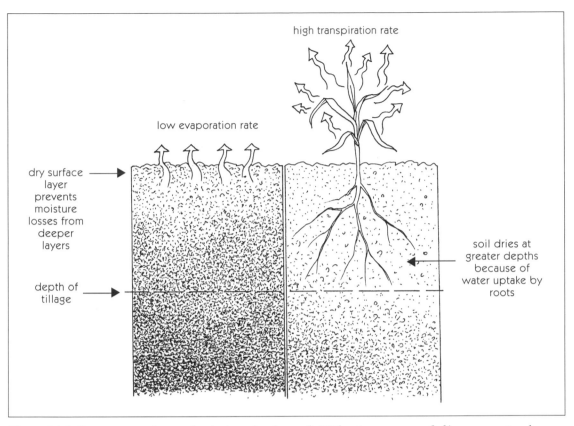

high transpiration rate

low evaporation rate

dry surface layer prevents moisture losses from deeper layers →

depth of tillage →

← soil dries at greater depths because of water uptake by roots

Figure 14.2 Cover crops enhance the drying of a clay soil. Without cover crops (left), evaporation losses are low after the surface dries. With cover crops (right), water is removed from deeper in the soil, because of root uptake and transpiration from plant leaves, resulting in better tillage and traffic conditions.

nation of cover crops with shallow and deep rooting systems are preferred. Ideally, such crops are part of the rotational cropping system, which is typically done on ruminant livestock-based farms.

The relative benefits of incorporating or mulching a cover/rotation crop are site-specific. Incorporation through tillage loosens the soil, which may be beneficial if the soil has been heavily trafficked. This would be the case with a sod crop that was actively managed for forage production, sometimes with traffic under rela-

tively wet conditions. Incorporation through tillage also encourages rapid nitrogen mineralization. Compared to plowing down a sod crop, cutting and mulching in a no-till or zone-till system reduces nutrient availability and does not loosen the soil. On the other hand, a heavy protective mat at the soil surface provides some weed control and better water infiltration and retention. Some farmers have been successful with cut-and-mulch systems involving aggressive, tall cover/rotation crops, such as rye and sorghum-sudangrass.

Addition of other organic materials. Adding animal manure, compost or sewage sludge benefits the surface soil layer in which they are incorporated by providing a source of organic matter. The long-term benefits of applying these materials, relative to soil compaction, may be very favorable, but in many cases, the spreaders themselves are a major cause of compaction. Livestock-based farms in humid regions usually apply manure using heavy spreaders (often with poor load distribution) on wet or marginally dry soils, resulting in severe compaction of both the surface layer and the subsoil. The need to incorporate animal manure for efficient nitrogen use and odor control is also often a barrier to the adoption of no-till or zone-till systems. This problem can be overcome only through an additional investment in manure injection tools. In general, the addition of organic materials should be done with care to obtain the biological and chemical benefits, while not aggravating compaction problems.

Controlled traffic and permanent beds. One of the most promising, but rarely adopted, practices for reducing soil compaction is the use of controlled traffic lanes. In this system, traffic associated with all field operations is limited to the same lanes. A controlled traffic system is easiest adopted with row crops in zone, ridge or no-till systems (not requiring full-field tillage, see chapter 15), because crop rows and traffic lanes remain recognizable year after year. Ridge tillage, in fact, dictates controlled traffic, as wheels cannot cross the ridges. Zone and no-till do not necessarily require controlled traffic, but greatly benefit from it, because the soil is not regularly loosened by aggressive tillage. Adoption of controlled traffic lanes typically requires some adjustment of field equipment to

insure that all wheel traffic occurs in the same lanes and also requires considerable discipline from equipment operators.

The primary benefit of controlled traffic is the lack of compaction for most of the field at the expense of limited areas that receive all the compaction. Because the degree of soil compaction doesn't necessarily worsen with each load (most of the compaction occurs with the heaviest loading and does not greatly increase with other passes), damage in the traffic lanes is not much more severe than that occurring on the whole field in a system with uncontrolled traffic. Controlled traffic lanes may actually have an advantage in that the consolidated soil is able to bear greater loads, thereby better facilitating

---

Crops Particularly Hard on Soils

✓ Potatoes require intensive tillage and return low rates of residue to soil.
✓ Silage corn and soybeans return low rates of residue.
✓ Many vegetable crops require timely harvest, so field traffic occurs even when the soils are too wet.

Special care is needed to counter the negative effects of such crops. These may include selecting soil-improving crops to fill out the rotation, extensive use of cover crops, using controlled traffic, and adding extra organic materials, such as manures and composts. In an 11-year experiment in Vermont with continuous corn silage on a clay soil, we found that applications of dairy manure were critical to maintaining good soil structure. Applications of 0, 10, 20, and 30 tons (wet weight) of dairy manure per acre each year of the experiment resulted in pore space of 44, 45, 47, and 50 percent of the soil volume.

BUILDING SOILS FOR BETTER CROPS

Using a Penetrometer for Assessing Compaction

A penetrometer is a tool that costs about $200 and measures the resistance to soil penetration. A penetrometer has a rod with a cone-shaped tip that is pushed into the soil. When penetration resistance is greater than about 300 psi, the soil is usually too hard for roots to grow (see chapter 6). Remember that the strength of the soil depends on the water content as well as bulk density (also chapter 6), so penetrometer measurements need to be repeated several times during the growing season to make a good assessment. However, you can sometimes get important information from a single set of penetrometer measurements made when the soil is very moist (for example at the beginning of the growing season in humid regions). If penetrometer readings at that time are near or above 300 psi, they will surely be higher when the soil dries out later in the season. When making penetrometer measurements, try to notice soil strength in both the plow layer and the subsoil — corrective action may be different for each case.

When using a penetrometer, also keep in mind that soil strength is extremely variable and multiple penetrations throughout the field should be made and averaged before drawing conclusions. Penetrometers do not work well in rocky soils, as the measurement is not valid when the tip hits a rock. The devices are not very good for predicting rooting behavior in clayey soils. Although clays may get very hard upon drying, they may still have enough large pores to allow roots to proliferate.

field traffic. Compaction also can be reduced significantly by maximizing traffic of farm trucks along the field boundaries and using planned access roads, rather than allowing them to randomly travel over the field.

...most of the compaction occurs
with the heaviest loading and
does not greatly increase
with other passes...

A permanent (raised) bed system is another way of controlling traffic. In this case, controlled traffic is combined with soil shaping to improve the physical conditions in the beds. Beds do not receive traffic after they've been formed. This is especially attractive where traffic on wet soil is unavoidable for economic reasons (for example, with certain fresh-market vegetable crops) and where it is useful to install equipment, such as irrigation lines, for multiple years.

SOURCES

Ontario Ministry of Agriculture, Food, and Rural Affairs. 1997. *Soil Management.* Best Management Practices Series. Available from the Ontario Federation of Agriculture, Toronto, Ontario (Canada).

Kok, H., R.K. Taylor, R.E. Lamond, and S. Kessen. 1996. *Soil Compaction: Problems and Solutions.* Cooperative Extension Service publication AF 115. Kansas State University. Manhattan, KS.

15

Reducing Tillage

*...the crying need is for a soil surface similar to
that which we find in nature. ...[and] the way to attain
it is to use an implement that is incapable of burying
the trash it encounters; in other words,
any implement except the plow.*

—E.H.FAULKNER, 1943

Although tillage is an ancient practice, the question of which tillage system is most appropriate for any particular field or farm is still difficult to answer. Before we discuss different tillage systems, let's consider why people started tilling ground. Intensive, full-field tillage was first practiced by farmers who grew small-grain crops, such as wheat, rye and barley, primarily in Western Asia, Europe, and Northern Africa. Tillage was needed to control weeds and give the crop a head-start before a new flush of weeds germinated. It also stimulated mineralization of organic forms of nitrogen to forms that plants could use. Mostly, however, intensive tillage created a fine seedbed, thereby greatly improving germination. The soil was typically loosened by plowing and then dragged to pulverize the clods and create a finely aggregated and smooth seedbed. The loosened soil also tended to provide a more favorable rooting environment, facilitating seedling survival and plant growth. From early on, animal traction was employed to accomplish this arduous task. At the end of the growing season, the entire crop was harvested, because the straw also had considerable economic value for animal bedding, roofing thatch, and brick making. Sometimes, fields were burned after crop harvest to remove remaining crop residues and to control pests. Although this tillage-cropping system lasted for a long time, it resulted in excessive erosion, especially in the Mediterranean region, where it caused extensive soil degradation. Eventually deserts spread as the climate became drier.

Other ancient agricultural systems, notably those in the Americas, did not use intensive full-field tillage for grain production. Instead, they used a hoe for manual tillage that created small mounds (hilling). This was well adapted to the regional staples of corn and beans, which have larger seeds and require lower planting densities than wheat, rye, and barley. Several seeds were placed in a small hill, often with the help of a planting stick, and hills were spaced several feet apart. In many, but not all cases, the hills were elevated to provide a temperature and moisture advantage to the crop. Compared with the cereal-based systems growing only one crop in a monoculture, these fields often included two or three plant species growing at the same time. This hilling system was generally less prone to erosion than whole-field tillage, but climate and soil conditions on steep slopes still frequently caused considerable soil degradation.

A third tillage system was practiced as part of the rice-growing cultures in southern and eastern Asia. Here, paddies were tilled to control weeds, puddle the soil, and create a dense layer that limited the downward losses of water through the soil. The puddling process occurred when the soil was worked while wet — in the plastic or liquid consistency state — and was specifically aimed at destroying soil structure. This system was designed because rice plants thrive under flooded conditions. There is little soil erosion, because paddy rice must be grown either on flat or terraced lands and runoff is controlled as part of the process of growing the crop.

Full-field tillage systems became more widespread as the influence of European culture expanded into other regions of the world. It's better adapted to mechanized agriculture so the traditional "hill crops" eventually became row crops. The invention of the moldboard plow provided a more effective tool for weed control by fully turning under crop residues, growing weeds, and weed seeds. The development of increasingly powerful and comfortable tractors made tillage an easier task. In fact, it has become almost a recreational activity for some farmers.

New technologies have lessened the need for tillage. The development of herbicides reduced the need for tillage as a weed control method. New planters achieved better seed placement, even without preparing a seedbed beforehand. Amendments, such as fertilizers and liquid manures, can be directly injected or band-applied. Now, there are even vegetable transplanters that provide good soil-root contact in reduced or no-till systems. Although herbicides often are used to kill cover crops before planting the main crop, farmers and researchers have found that they can obtain fairly good cover crop control through well-timed mowing, rolling, or rolling/chopping — greatly reducing the amount of herbicide needed.

Technologies have lessened
the need for tillage...

→ herbicides

→ new planters and transplanters

→ new physical methods for cover crop suppression

Increased mechanization, intensive tillage, and erosion have degraded many agricultural soils to such an extent that people think they require tillage to provide temporary relief from compaction. As aggregates are destroyed, crusting and compaction create a soil "addicted" to tillage. Except perhaps for organic crop production systems, where tillage is needed because herbicides aren't used, a crop can be produced with limited or no tillage with the same economic return as conventional tillage systems. Managing soil in the right way to make reduced tillage systems successful, however, remains a considerable challenge.

TILLAGE SYSTEMS

Tillage systems are often classified by the amount of surface residue left on the soil surface. Conservation tillage systems are those that leave more than 30 percent of the soil surface covered with crop residue. This is considered to be a level at which erosion is significantly reduced (see figure 15.1). Of course, this partially depends on the amount of residue left after harvest, which may vary greatly among crops and harvest method (for example, corn harvested for grain or silage). Although surface residue cover greatly influences erosion potential, the sole focus on it is somewhat misleading. Erosion potential also is affected by factors such as surface roughness and soil loosening. Another distinction of tillage systems is whether they are *full-field* systems or *restricted* tillage systems (figure 15.2). The benefits and limitations of various tillage systems are compared in table 15.1.

Conventional Tillage

A full-field system manages the soil uniformly across the entire field surface. It typically involves a primary pass to loosen the soil and incorporate materials at the surface (fertilizers, amendments, weeds, etc.), followed by one or more secondary tillage passes to create a suitable seedbed. Primary tillage tools are generally moldboard plows,

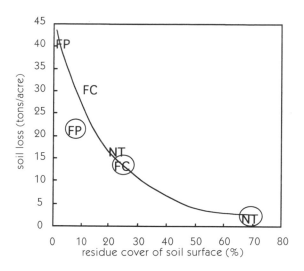

This figure shows that:
- Surface residue reduces erosion.
- Reduced tillage (chisel and no-till) leaves more residue and results in less erosion than plowing.
- Corn (circled) returns more residue than soybeans.

Figure 15.1 Soil erosion dramatically decreases with increasing surface cover. (Fall plow (FP), fall chisel (FC), no-till (NT), corn = circles, soybeans = no circles). Modified from Manuring, 1979.

chisels, and disks, while secondary tillage is accomplished with finishing disks, tine or tooth harrows, rollers, packers, drags, etc. These tillage systems create a uniform and often finely aggregated seedbed over the entire surface of the field. Such systems appear to perform well because they create near-ideal conditions for seed germination and crop establishment.

Moldboard plowing is generally the least desirable practice because it is energy intensive, leaves very little residue on the surface, and often requires multiple secondary tillage passes. It also tends to create plow pans. However, it is generally the most reliable practice and almost always results in reasonable crop growth. Chisel implements generally provide results similar to the moldboard plow, but require less energy and leave significantly more residue on the surface. Chisels also allow for more flexibility in depth of tillage, generally from 5 to 12 inches, with some tools specifically designed to go deeper. Disks usually perform shallow tillage, depending on their size, and still leave residue on the surface. They can be used as both primary and secondary tillage tools.

Although full-field tillage systems have their disadvantages, they often can help overcome certain problems, such as compaction and high weed pressures. Organic farmers often use moldboard plowing as a necessity to provide adequate weed control and facilitate nitrogen release from incorporated legumes. Livestock-based farms often use a plow to incorporate manure and to help make rotation transitions from sod crops to row crops.

Figure 15.2 Four tillage systems.
a) Chisel tillage: Shanks provide full-field soil loosening.
b) No-till: Corn was directly planted into untilled soil. Photo by NRCS.
c) Zone tillage: Planter loosens soil in the row and moves residues to the side.
d) Ridge tillage: Crop is planted into small ridges without tillage.

BUILDING SOILS FOR BETTER CROPS

TABLE 15.1
Tillage System Benefits and Limitations

TILLAGE SYSTEM	BENEFITS	LIMITATIONS
FULL-FIELD TILLAGE		
Moldboard plow	Easy incorporation of fertilizers and amendments. Buries surface weed seeds. Soil dries out fast. Temporarily reduces compaction.	Leaves soil bare. Destroys natural aggregation and enhances organic matter loss. Surface crusting and accelerated erosion common. Causes plow pans. High energy requirements.
Chisel plow	Same as above, but with more surface residues.	Same as above, but less aggressive destruction of soil structure, less erosion, less crusting, no plow pans, and less energy use.
Disc harrow	Same as above.	Same as above
RESTRICTED TILLAGE		
No-till	Little soil disturbance. Few trips over field. Low energy use. Most surface residue cover and erosion protection.	Hard to incorporate fertilizers and amendments. Wet soils slow to dry and warm up in spring. Can't alleviate compaction by using tillage.
Zone-till	Same as above.	Same as above, but fewer problems with compaction.
Ridge-till	Easy incorporation of fertilizers and amendments. Some weed control as ridges are built. Seed zone on ridge dries and warms more quickly.	Hard to use together with sod-type or narrow-row crop in rotation. Equipment needs to be adjusted to travel without disturbing ridges.

> A good planter is perhaps the most important secondary tillage tool, because it helps overcome poor soil-seed contact without destroying surface aggregates over the entire field.

Besides incorporating surface residue, full-field tillage systems with intensive secondary tillage crush the natural soil aggregates. The pulverized soil does not take heavy rainfall well. The lack of surface residue causes sealing at the surface, which generates runoff and erosion and creates hard crusts after drying. Also, intensively tilled soil will settle after some rainfall and may "hardset" upon drying, thereby restricting root growth.

Full-field tillage systems can be improved by using tools, such as chisels (figure 15.2a), that leave some residue on the surface. Reducing secondary tillage also helps decrease negative aspects of full-field tillage. Compacted soils tend to till up cloddy and intensive harrowing and packing is then seen as necessary to create a good seedbed. This creates a vicious cycle of further soil degradation with intensive tillage. Secondary tillage often can be reduced through the use of state-of-the-art planters, which create a finely aggregated zone around the seed without requiring the entire soil width to be pulverized. Indeed, a good planter is perhaps the most important secondary tillage tool, because it helps overcome poor soil-seed contact without destroying surface aggregates over the entire field. A fringe benefit of reduced secondary tillage is

that rougher soil has much higher water infiltration rates and reduces problems with settling and hardsetting after rains. Weed seed germination is also generally reduced, but pre-emergence herbicides tend to be less effective than with smooth seedbeds. Reducing secondary tillage may, therefore, require emphasis on post-emergence weed control.

Restricted Tillage Systems

These systems are based on the idea that tillage can be limited to the area around the plant row and does not have to disturb the entire field. Three tillage systems fit this concept — no-till, zone-till, and ridge-till.

The no-till system (figure 15.2b) does not involve any soil loosening, except for a very narrow and shallow area immediately around the seed zone. This localized disturbance is typically accomplished with a fluted, or ripple, coulter on a planter. This is the most extreme change from conventional tillage.

> Restricted tillage systems are based on the idea that tillage can be limited to the area around the plant row and does not have to disturb the entire field.

No-till systems have been used successfully on many soils in different climates. They are especially well adapted to coarse-textured soils (sands and gravels), as they tend to be softer and less susceptible to compaction. It typically takes a few years for no-tilled soils to improve, after which no-till crops often out-yield crops grown with conventional tillage. The quality of no-tilled soils, by almost any measure, improves over time. The maintenance of surface residue protects against erosion and increases biological activity by protecting the soil from temperature and heat extremes. Surface residues also reduce water evaporation, which — combined with deeper rooting — reduces the susceptibility to drought.

Another system, usually called zone tillage (figure 15.2c), gets some of the benefits of soil disturbance in the soil around the plant row without disturbing the entire field. It uses multiple fluted coulters mounted on the front of a planter (figure 15.3) to develop a fine seedbed of approximately 6 inches wide by 4 inches deep, and typically uses trash wheels to move residue away from the row. The system may also include a separate pass of a "zone building" implement during the off season (see figure

Figure 15.3 Zone-till planter. a. Coulters (cut up residues and break up soil in seed zone); b. Fertilizer disc openers (place granular starter fertilizer in a band next to the seed); c. Spider (trash) wheels (move residue away from the row); d. Seed placement unit; e. Press wheels (create firm seedbed); and f. Wheel used for transporting the planter.

14.1). This typically involves a narrow shank or knife, sometimes used to inject fertilizers, combined with a trash remover or hilling disk (to perform in-row tillage and overcome compaction problems). The term "strip tillage" often is used to describe the latter system.

Ridge tillage (figure 15.2d) combines some tillage with a ridging operation. This system is particularly attractive for cold and wet soils, because the ridges offer developing plants a warmer and better-drained environment. The ridging operation can be combined with mechanical weed control and allows for band application of herbicides. This decreases the cost of chemical weed control, allowing for about a two-thirds reduction in herbicide use.

For fine and medium-textured soils, zone and ridge tillage systems generally work better than strict no-till, especially in regions with cold and wet springs. Because these soils are more susceptible to compaction, some soil disturbance is probably beneficial. No-till is used successfully for narrow-row crops, including small grains, perennial legumes and grasses. Zone and ridge tillage are only adapted to wide-row crops with 30-inch spacing or more.

WHICH TILLAGE SYSTEM FOR YOUR FARM?

This is difficult to answer. It depends on your climate, soils, cropping systems, and your objectives. Here are some general guidelines.

Grain and vegetable farms have great flexibility with adopting reduced tillage systems. In the long run, limited disturbance and residue cover improve soil quality, reduce erosion, and boost yields. A negative aspect of these systems is that, at first, they may require more herbi-

Figure 15.4 No-till transplanting of vegetables into a killed cover crop on the Groff farm in Pennsylvania. Photo by Ray Weil.

cides. However, combining reduced tillage with use of cover crops frequently helps reduce weed problems. Weed pressures typically decrease significantly after a few years. Mulched cover crops, as well as newly designed mechanical cultivators, help provide effective weed control in high-residue systems. Some innovative farmers use no-till combined with a cover crop, which is mowed or otherwise killed to create a thick mulch. Steve Groff, a vegetable and field crop farmer in Pennsylvania, modified a rolling stalk chopper to roll down and crimp his vetch/rye cover crop, providing weed control with minimal use of herbicides (see profile, p. 145).

Farmers need to be aware of potential soil compaction problems with reduced tillage. If a strict no-till system is used on a compacted soil, especially on medium or fine-textured soils, serious yield reductions may occur. As discussed in chapter 6, dense soils have a narrow water range for which plant growth is not restricted. Crops growing on compacted soils are more susceptible to inadequate aeration during wet periods and restricted root growth and inadequate moisture during drier periods. Compaction,

therefore, reduces plant growth and makes crops more susceptible to pest pressures.

In poorly structured soils, tools like zone-builders and zone-till planters may provide compaction relief in the row, while maintaining an undisturbed soil surface. Over time, soil structure improves, unless recompaction occurs from other field operations. Crops grown on imperfectly drained soils tend to benefit greatly from ridging or bedding, because part of the root zone remains aerobic during wet periods. These systems also use controlled traffic lanes, which greatly reduce compaction problems. Unfortunately, matching wheel spacing and tire width for planting and harvesting equipment is not always an easy task.

Before Converting to No-Till

An Ohio farmer asked one of the authors what could be done about a compacted, low organic matter, and low fertility field that had been converted to no-till a few years before.

Clearly, the soil's organic matter and nutrient levels should have been increased and the compaction alleviated before the change. Once you're committed to no-till, you've lost the opportunity to easily and rapidly change the soil's fertility or physical properties. The recommendation is really the same as for someone establishing a perennial crop like an apple orchard. Build up the soil and remedy compaction problems before converting to no-till. It's going to be much harder to do later on.

Once in no-till, there are some things that can be tried to break up compacted soils, such as a sorghum-sudangrass cover crop. However, a severe compaction problem may require tilling up the soil and starting over.

For organic farms, as with traditional farms before agrichemicals were available, full-field tillage may be necessary for mechanical weed control and incorporation of manures and composts. Organic farming on lands prone to erosion may, therefore, involve tradeoffs. Erosion can be reduced by using rotations with perennial crops and a modern planter to establish good crop stands without excessive secondary tillage. In addition, soil structure may be easier to maintain, because organic farms generally use more organic inputs, such as manures and composts.

Livestock-based farms face special challenges related to applying manures or composts to the soil. Although these materials may sometimes be injected directly, some type of tillage usually is needed to avoid large losses of nitrogen by volatilization and phosphorus and pathogens by runoff. Transitions from sod to row crops are usually easier with some tillage. Farmers raising livestock should try to reduce tillage as much as possible and use methods that leave residue on the surface.

ROTATE TILLAGE SYSTEMS?

A tillage program does not need to be rigid. When changing to reduced tillage, consider incorporating nutrients and organic matter with the moldboard plow (see box on p. 142). Fields that are zone or no-tilled may occasionally need a full-field tillage pass to provide compaction relief or to incorporate amendments. Tilling a no-till or zone-till field should be done only if clearly needed. Although a flexible tillage program offers a number of benefits, aggressive tillage with a moldboard plow and harrows on soils for which no-till is best adapted will destroy the favorable soil structure built up by years of no-till management.

TIMING OF FIELD OPERATIONS

The success of a tillage system depends on many other factors. For example, reduced tillage systems, especially in the early transition years, may require more attention to nitrogen management, as well as weed, insect, and disease control. Also, the performance of tillage systems may be affected by the timing of field operations. If tillage or planting is done when the soil is too wet (its water content is above the plastic limit) then cloddiness and poor seed placement may result in poor stands. Tillage also is not recommended when soil is too dry, because of excess dust creation, especially on compacted soils. A "ball test" (Chapter 6) helps ensure that field conditions are right.

Frost Tillage?

You may have heard of frost seeding legumes into a pasture, hayfield, or winter wheat crop in very early spring, but probably never heard of tilling a frozen soil. It seems a strange concept, but some farmers are using frost tillage as a way to be timely and reduce unintended tillage damage. It can be done after frost has first entered the soil, but before it has penetrated more than 4 inches. Water moves upward to the freezing front and the soil underneath dries. This makes it tillable as long as the frost layer is not too thick. Compaction is reduced because equipment is supported by the frozen layer. The resulting rough surface is favorable for water infiltration and runoff prevention. Some livestock farmers like frost tillage as a way to incorporate or inject manure in the winter.

Optimum Tillage System

New agricultural technologies provide opportunities to reduce tillage and improve soil quality. The optimum system for any farm depends mainly on soils and climate, as well as the need for mechanical weed control, incorporation of cover crops and animal manures, and lessening compaction. Tillage systems should change in the direction of those that leave residue and mulches on the surface and that limit the pulverization of soil aggregates.

Because soil compaction may affect the success of a reduced tillage system, a whole-system approach to soil management is needed. For example, no-till systems that also involve harvesting operations with heavy equipment will succeed only if traffic can be restricted to dry conditions or fixed lanes within the field. Even zone-tillage methods may fail if heavy harvest equipment is used without fixed lanes. Soils that are severely eroded and low in organic matter may need careful management when making the transition to reduced tillage systems. In such cases, methods that increase the soil organic matter content and improve soil structure (for example, cover cropping and organic amendments) before reducing tillage will improve the probability for success of these systems. As surface residue levels increase with the start of reduced tillage, some soil loosening may be needed to relieve compaction.

SOURCES

Cornell Recommendations for Integrated Field Crop Production. 2000. Cornell Cooperative Extension, Ithaca, NY.

Manuring. 1979. Cooperative Extension Service Publication AY-222, Purdue University. West Lafayette, IN.

Ontario Ministry of Agriculture, Food, and Rural Affairs. 1997. *No-till: Making it Work.* Available from the Ontario Federation of Agriculture, Toronto, Ontario (Canada).

van Es, H.M., A.T. DeGaetano, and D.S. Wilks. 1998. Upscaling plot-based research information: Frost tillage. *Nutrient Cycling in Agroecosystems.* 50: 85–90.

Steve Groff
Lancaster County, Pennsylvania

Steve Groff raises grains and vegetables every year on his 175-acre farm in Lancaster County, Pa., but his soil shows none of the degradation that can occur with intensive cropping. Mixing cash crops such as corn, alfalfa, soybeans, broccoli, tomatoes and peppers with cover crops and a unique no-till system, Groff has kept portions of his farm untouched by a plow for close to two decades.

"No-till is a practical answer to concerns about erosion, soil quality, and soil health," says Groff, who won a national no-till award in 1999. "I want to leave the soil in better condition than I found it."

Groff confronted a rolling landscape pocked by gullies when he began farming with his father after graduating from high school. They regularly used herbicides and insecticides, tilled annually or semi-annually and rarely used cover crops. Like other farmers in Lancaster County, they ignored the effects of tillage on a sloped landscape that causes an average of 9 tons of soil per acre to wash into the Chesapeake Bay every year.

Tired of watching two-feet-deep crevices form on the hillsides after every heavy rain, Groff began experimenting with no-till to protect and improve the soil.

"We used to have to fill in ditches to get machinery in to harvest," Groff says. "I didn't think that was right."

However, Groff stresses that switching to no-till alone isn't enough. He has created a new system, reliant on cover crops, rotations, *and* no-till, to improve the soil. He's convinced such methods contribute to better yields of healthy crops, especially during weather extremes.

He pioneered what he likes to call the "Permanent Cover" cropping system when the Pennsyl-vania chapter of the Soil and Water Conservation Society bought a no-till transplanter for vegetable crops. Groff was one of the first farmers to try it. The machine allows him to transplant seedlings into slots cut into cover crop residue. The slots are just big enough for the young plants and do not disturb the soil on either side. The result: Groff can prolong the erosion-slowing benefits of cover crops.

Groff's no-till system relies on winter cover crops and residues that blanket the soil virtually all year. In the fall, he uses a no-till seeder to plant a combination of rye and hairy vetch. Groff likes the pairing because of their complementary benefits. Their root structures grow in different patterns, and the vegetation left behind after killing leaves different residues on the soil surface.

Groff uses a rolling stalk-chopper — modified from Midwest machines that chop corn stalks after harvest — to kill the covers in the spring. The chopper flattens and crimps the cover crop, providing a thick mulch. Once it's flat, he makes a pass with the no-till planter or transplanter.

The system creates a very real, side benefit in reduced insect pest pressure. Once an annual problem, Colorado potato beetle damage has all but disappeared from Groff's tomatoes. Since he began planting into the mulch, he has greatly reduced spraying pesticides. The thick mat also prevents splashing of soil during rain, a primary cause of early blight on tomatoes.

"We have slashed our pesticide and fertilizer bill nearly in half, compared to a conventional tillage system," Groff says. "At the same time, we're building valuable topsoil and not sacrificing yields."

"No-till is not a miracle, but it works for me," he says. "It's good for my bottom line, I'm saving soil, and I'm reducing pesticides and increasing profits."

Groff is convinced his crops are better than those produced in soils managed conventionally, especially during weather extremes. He has noted high earthworm populations and other biological activity deep in the soil.

Ray Weil, a soils professor at the University of Maryland who has spent time on Groff's farm, concurs.

"Steve's subsoil is like other farmers' topsoil," he says.

Groff promotes his system at annual summer field days that draw huge crowds of farmers and at his innovative web site, <www.cedarmeadowfarm .com>.

16

Nutrient Management: An Introduction

The purchase of plant food is an important matter,
but the use of a [fertilizer] is not a cure-all, nor will it
prove an adequate substitute for proper soil handling.

—J.L. HILLS, C.H. JONES, AND C. CUTLER, 1908

Of the 18 elements needed by plants, only three — nitrogen (N), phosphorus (P), and potassium (K) — are commonly deficient in soils. Deficiencies of other nutrients, such as magnesium, sulfur, zinc, boron, and manganese, certainly occur, but they are not as widespread. However, in locations with lots of young minerals that haven't been weathered much by nature, such as the Dakotas, potassium deficiencies are less common. Deficiencies of sulfur, magnesium, and some micronutrients may be more widespread in regions with highly weathered minerals, such as the southeastern states, or those with high rainfall, such as portions of the Pacific Northwest.

Environmental concerns have placed more emphasis on better management of nitrogen and phosphorus over the last few decades. These nutrients are critical to soil fertility management, but they are also responsible for widespread environmental problems. Poor soil and crop management, the overuse of fertilizers, misuse of manures, sludges and composts, and high animal numbers on limited land area have contributed to surface and groundwater pollution in many regions of the U.S. Because both nitrogen and phosphorus are used in large quantities and their overuse has potential environmental implications, we'll discuss them together in chapter 17. Other nutrients, cation exchange, soil acidity (low pH) and liming, and arid and semi-arid region problems with sodium, alkalinity (high pH), and excess salts are covered in chapter 18.

THE BOTTOM LINE: NUTRIENTS AND PLANT HEALTH, PESTS, PROFITS, AND THE ENVIRONMENT

Management practices are all related. The key is to visualize them all as whole-farm management, leading you to the goals of better crop growth and better environmental quality. If a soil has good tilth, no subsurface compaction, good drainage, adequate water, and a good supply of organic matter, then plants should be healthy and have large root systems. A large and healthy root system enables plants to efficiently take up nutrients and water from the soil and to use those nutrients to produce higher yields.

Doing a good job of managing nutrients on the farm and in individual fields is critical to general plant health and management of plant pests. Too much available nitrogen in the early part of the growing season allows small-seeded weeds, with few nutrient reserves, to get well established. This early jump-start may then enable them to out-compete crop plants later on. Crops do not grow properly if nutrients aren't present in sufficient quantities and in reasonable balance to one another. Plants may be stunted if nutrients are low, or they may grow too much foliage and not enough fruit if nitrogen is too plentiful relative to other nutrients. Plants under nutrient stress, such as too low or too high nitrogen levels, are not able to emit as much of the natural chemicals that signal beneficial insects when insect pests feed on leaves or fruit. Stalk rot of corn is aggravated by low potassium levels. On the other hand, pod rot of peanuts is associated with excess potassium within the fruiting zone of peanuts (the top 2 to 3 inches of soil). Blossom-end rot of tomatoes is related to low calcium levels, often brought

---

The ABCs of Nutrient Management

a. Build up and maintain high soil organic matter levels.

b. Test manures and credit their nutrient content before applying fertilizers or other amendments.

c. Incorporate manures into the soil quickly, if possible, to reduce nitrogen volatilization and potential loss of nutrients in runoff.

d. Test soils regularly to determine the nutrient status and whether or not manures, fertilizers, or lime are needed.

e. Balance nutrient inflows and outflows to maintain optimal levels and allow a little "draw-down" if nutrient levels get too high.

f. Enhance soil structure and reduce field runoff by minimizing soil compaction damage.

g. Use forage legumes or legume cover crops to provide nitrogen to following crops and develop good soil tilth.

h. Use cover crops to tie up nutrients in off-season, enhance soil structure, and reduce runoff and erosion.

i. Maintain soil pH in the optimal range for the crops in your rotation.

j. When phosphorus and potassium are very deficient, broadcast some of the fertilizer to increase the general soil fertility level, and band apply some as well.

k. To get the most efficient use of the fertilizer when phosphorus and potassium levels are in the medium range, consider band application at planting, especially in cool climates.

Essential Nutrients for Plants

ELEMENT	COMMON AVAILABLE FORM	SOURCE
needed in large amounts		
Carbon	CO_2	atmosphere
Oxygen	O_2, H_2O	atmosphere and soil pores
Hydrogen	H_2O	water in soil pores
Nitrogen	NO_3^-, NH_4^+	soil
Phosphorus	$H_2PO_4^-$, HPO_4^{-2}	soil
Potassium	K^+	soil
Calcium	Ca^{+2}	soil
Magnesium	Mg^{+2}	soil
Sulfur	SO_4^{-2}	soil
needed in small amounts		
Iron	Fe^{+2}, Fe^{+3}	soil
Manganese	Mn^{+2}	soil
Copper	Cu^+, Cu^{+2}	soil
Zinc	Zn^{+2}	soil
Boron	H_3BO_3	soil
Molybdenum	MoO_4^{-2}	soil
Chlorine	Cl^-	soil
Cobalt	Co^{+2}	soil
Nickel	Ni^{+2}	soil

on by droughty, or irregular rainfall/irrigation, conditions.

When plants either don't grow well or are more susceptible to pests, this affects the economic return. Yield and crop quality usually are reduced, lowering the amount of money received. There also may be added costs to control pests that take advantage of poor nutrient management. In addition, when nutrients are applied beyond plant needs, it's like throwing money away. And when nitrogen and phosphorus are lost from the soil by leaching to groundwater or running into surface water, entire communities may suffer from poor water quality.

ORGANIC MATTER AND NUTRIENT AVAILABILITY

The best single overall strategy for nutrient management is to work to enhance soil organic matter levels in soils. This is especially true for nitrogen and phosphorus. Soil organic matter, together with any freshly applied residues, are well known sources of available nitrogen for plants. Mineralization of phosphorus and sulfur from organic matter is also an important source of these nutrients. As discussed earlier, organic matter helps hold on to potassium (K^+), calcium (Ca^{++}), and magnesium (Mg^{++}) ions. It also provides natural chelates that maintain micronutrients such as zinc, copper, and manganese, in forms that plants can use.

IMPROVING NUTRIENT CYCLING ON THE FARM

For economic and environmental reasons, it makes sense to utilize nutrient cycles efficiently. Goals should include the reduction in long-distance nutrient flows, as well as promoting "true" on-farm cycling. There are a number of strategies to help farmers reach the goal of better nutrient cycling:

- **Reduce unintended losses** by promoting water infiltration and better root health through enhanced management of soil organic matter and physical properties. The ways in which organic matter can be built up and maintained include increased additions of a variety of sources of organic matter plus methods for reducing losses via tillage and conservation practices.
- **Enhance nutrient uptake efficiency** by carefully using fertilizers and amendments. Better placement and synchronizing application with plant growth both improve efficiency. Sometimes, changing planting dates or switching to a new crop creates a better match between the timing of nutrient availability and crop needs.
- **Tap local nutrient sources** by seeking local sources of organic materials, such as leaves or grass clippings from towns, aquatic weeds harvested from lakes, produce waste from markets and restaurants, food processing wastes, and clean sewage sludges (see discussion on sewage sludge in chapter 8). Although some of these do not contribute to true nutrient cycles, the removal of agriculturally usable nutrients from the "waste stream" makes sense and helps develop more environmentally sound nutrient flows.

- **Promote consumption of locally produced foods** by supporting local markets as well as returning local food wastes to farmland. When people purchase locally produced foods there are more possibilities for true nutrient cycling to occur. Some Community Supported Agriculture farms, where subscriptions are paid before the start of the growing season, encourage their members to return produce waste to the farm for composting, completing a true cycle.
- **Reduce exports of nutrients in farm products** by adding animal enterprises to crop farms (fewer exports, and more reason to include forage legumes and grasses in rotation). The best way to both reduce nutrient exports per acre, as well as to make more use of forage legumes in rotations, is to add an animal (especially a ruminant) enterprise to a crop farm. Compared with selling crops, feeding crops to animals and exporting animal products results in far fewer nutrients leaving the farm. (Keep in

---

Strategies for Improving Nutrient Cycles

✓ Reduce unintended losses.
✓ Enhance nutrient uptake efficiency.
✓ Tap local nutrient sources.
✓ Promote consumption of locally produced foods.
✓ Reduce exports of nutrients in farm products.
✓ Bring animal densities in line with the land base of the farm.
✓ Develop local partnerships to balance flows among different types of farms.

---

mind that, on the other hand, raising animals with mainly purchased feed is the best way to overload a farm with nutrients.)

- **Bringing animal densities in line with the land base of the farm** can be accomplished by renting or purchasing more land — to grow a higher percentage of animal feeds and for manure application — or by reducing animal numbers.
- **Develop local partnerships to balance flows among different types of farms.** As pointed out in chapter 8 when we discussed organic matter management, sometimes neighboring farmers cooperate with both nutrient management and crop rotations. This is especially beneficial when a livestock farmer has too many animals and imports a high percentage of feed and a neighboring vegetable farm has a need for nutrients and has an inadequate land base to allow a rotation that includes a forage legume. By cooperating with nutrient management and rotations, both farms win, sometimes in ways that were not anticipated (see "win-win" box). Encouragement and coordination from an extension agent may help neighboring farmers work out cooperative agreements. It is more of a challenge as the distances become greater.

Some livestock farms that are overloaded with nutrients are finding that composting is an attractive alternative way to handle manure. During the composting process, volume and weight are greatly reduced (see chapter 12), resulting in less material to transport. Organic farmers are always on the lookout for reasonably priced animal manures and composts. The landscape industry also uses a fair amount of compost. Local

---

Win-Win Cooperation

Cooperation between Maine potato farmers and their dairy farm neighbors has led to better soil and crop quality for both types of farms. As potato farmer John Dorman explains, after some years of cooperating with a dairy farm on rotations and manure management, soil health "...has really changed more in a few years than I'd have thought possible." Dairy farmer Bob Fogler feels that the cooperation with the potato farmer allowed his family to expand the dairy herd. He notes that "we see fewer pests and better quality corn. Our forage quality has improved. It's hard to put a value on it, but forage quality means more milk."

FROM HOARD'S DAIRYMAN, 4/10/99

---

or regional compost exchanges can help remove nutrients from overburdened animal operations and place them on nutrient-deficient soils.

USING FERTILIZERS AND AMENDMENTS

There are four main issues when applying nutrients:

- How much is needed?
- What source(s) should be used?
- When should the fertilizer or amendment be applied?
- How should the fertilizer or amendment be applied?

Chapter 19 details use of soil tests to help you decide how much fertilizer or organic nutrient sources to apply. Here we will go over how to approach the other three issues.

Nutrient Sources: Commercial Fertilizers vs. Organic Materials

There are numerous fertilizers and amendments normally used in agriculture (some are listed in table 16.1). Fertilizers such as urea, triple superphosphate, and muriate of potash (potassium chloride) are convenient to store and use. They are also easy to blend to meet nutrient needs in specific fields and provide predictable effects. Their behavior in soils and the ready availability of the nutrients is well established. The timing, rate, and uniformity of nutrient application is easy to control when using commercial fertilizers. However, there also are drawbacks to using commercial fertilizers. All of the commonly used nitrogen materials (those containing urea, ammonia, and ammonium) are acid forming, and their use in humid regions, where native lime has been weathered out, requires more frequent lime additions. Also, the high nutrient solubility can result in salt damage to seedlings when excess fertilizer is applied close to seeds or plants. Because nutrients in commercial fertilizers are readily available, under some circumstances more may leach to groundwater than when using organic nutrient sources. For example, high rainfall events on a sandy soil soon after ammonium nitrate fertilizer application will probably cause more nitrate loss than if a compost had been applied or a legume cover crop recently incorporated. Likewise, sediments lost by erosion from fields fertilized with commercial fertilizers probably will contain more available nutrients than those from fields fertilized with organic sources, resulting in more severe water pollution.

Organic sources of nutrients have many other good qualities too. They usually provide a more slow release source of fertility and the nitrogen availability is more evenly matched to the needs of growing plants. Sources like manures or crop residues commonly contain all the needed nutrients, including the micronutrients — but they may not be present in the proper proportion for soil and crop needs. These materials are also sources of soil organic matter, providing food for soil organisms and forming aggregates and humus.

One of the drawbacks to organic materials is the variable amounts and uncertain timing of nutrient release for plants to use. The value of manure as a nutrient source depends upon the type of animal, its diet, and how the manure is handled. For cover crops, the nitrogen contribution depends upon the species, amount of growth in the spring, and weather. Also, manures typically are bulky and may contain a high percentage of water — so considerable work is

---

Organic Farming vs. Organic Nutrient Sources

We've used the term "organic sources" of nutrients to refer to nutrients contained in crop residues, manures, and composts. These types of materials are used by all farmers—"conventional" or "organic." Both also use limestone and a few other materials. However, most of the commercial fertilizers listed in table 16.1 are not allowed in organic production. In place of sources such as urea, anhydrous ammonia, diammonium phosphate (DAP), concentrated superphosphate, and muriate of potash, organic farmers use products that come directly from minerals such as greensand, granite dust, and rock phosphate. Other organic products come from parts of organisms such as bone meal, fish meal, soybean meal, and bloodmeal (see table 16.2).

BUILDING SOILS FOR BETTER CROPS

TABLE 16.1
Composition of Various Common Amendments
and Commercial Fertilizers (%)

	N	P_2O_5	K_2O	Ca	Mg	S	Cl
N Materials							
Anhydrous ammonia	82						
Aqua ammonia	20						
Ammonium nitrate	34						
Ammonium sulfate	21					24	
Calcium nitrate	16			19	1		
Urea	46						
UAN solutions	28–32						
P and N+P Materials							
Superphosphate (ordinary)		20		20		12	
Triple superphospate		46		14		1	
Diammonium phosphate (DAP)	18	46					
Monoammonium phosphate (MAP)	11–13	48–52					
K Materials							
Potassium chloride (muriate of potash)			60				47
Potassium-magnesium sulfate ("Sul-Po-Mag")			22		11	23	2
Potassium sulfate			50		1	18	2
Other Materials							
Gypsum				23		17	
Limestone, calcitic				25–40	0.5–3		
Limestone, dolomitic				19–22	6–13		1
Magnesium sulfate				2	11	14	
Potassium nitrate	13		44				
Sulfur						30–99	
Wood ashes		2	6	23	2		

—OMAFRA, 1997.

needed to apply them per unit of nutrients. The timing of nutrient release is uncertain, because it depends both on the type of organic materials used and on the action of soil organisms. Their activities change with temperature and rainfall. Finally, the relative nutrient concentrations for a particular manure used may not match soil needs. For example, manures may contain high amounts of both nitrogen and phosphorus when your soil already has high phosphorus levels.

Selection of Commercial Fertilizer Sources

There are numerous forms of commercial fertilizers, many given in table 16.1. When you buy fertilizers in large quantities, the cheapest source is usually chosen. When you buy bulk blended fertilizer, you usually don't know what sources were used unless you ask. All you know is that it's a 10-20-20 (referring to the percent of available N, P_2O_5, and K_2O), or a 20-10-10, or an-

TABLE 16.2
Products Used by Organic Growers to Supply Nutrients*

	%N	%P_2O_5	%K_2O
Alfalfa pellets	2.7	0.5	2.8
Blood meal	13.0	2.0	-
Bone meal	3.0	20.0	0.5
Cocoa shells	1.0	1.0	3.0
Colloidal phosphate	-	18.0	-
Compost	1.0	0.4	3.0
Cottonseed meal	6.0	2.0	2.0
Fish scraps, dried & ground	9.0	7.0	-
Granite dust	-	-	5.0
Greensand	-	-	7.0
Hoof & horn meal	11.0	2.0	-
Linseed meal	5.0	2.0	1.0
Rock phosphate	-	30.0	-
Seaweed, ground	1.0	0.2	2.0
Soybean meal	6.0	1.4	4.0
Tankage	6.5	14.5	-

* 1. Values of P_2O_5 and K_2O represent *total* nutrients present. For fertilizers listed in table 16.1, the numbers are the amount that are *readily available*.

2. Organic growers also use potassium-magnesium sulfate ("Sul-Po-Mag" or "K-Mag"), wood ashes, limestone and gypsum listed in table 16.1. Although some use only manure that has been composted, others will use aged manures (see chapter 9). There are also a number of commercial organic products with a variety of trade names.

—PARNES, R. 1990.

other blend. However, below are a number of examples where you might not want to apply the cheapest source:

- Although the cheapest nitrogen form is anhydrous ammonia, the problems with injecting it into a soil with many large stones or the losses that might occur when injecting it into very moist clay may call for other nitrogen sources to be used instead.
- If both nitrogen and phosphorus are needed, diammonium phosphate (DAP) is a good choice because it has approximately the same cost and phosphorus content as concentrated superphosphate and also contains 18 percent N.
- Although muriate of potash (potassium chloride) is the cheapest potassium source, it may not be the best choice under certain circumstances. If you also need magnesium and don't need to lime the field, potassium-magnesium sulfate would be a better choice.

Method and Timing of Application

The timing of fertilizer application is frequently related to the application method chosen, so in this section we'll go over both issues together.

Broadcast application, where fertilizer is evenly distributed over the whole field and then usually incorporated during tillage, is best used to increase the nutrient level of the bulk of the soil. It is especially useful to build phosphorus and potassium when they are very deficient. Broadcasting for incorporation is usually done in the fall or in spring just before tillage. Broadcasting on top of a growing crop, called **topdress,** is commonly used to apply nitrogen, especially to crops that occupy the entire soil surface, such as wheat or a grass hay crop. [Amendments used

in large quantities, like lime and gypsum, are also broadcast prior to incorporation into the soil.]

There are various types of localized placement of fertilizer. **Banding** small amounts of fertilizer to the side and below the seed at planting is a common application method. It is especially useful for row crops grown in cool soil conditions, such as early in the season, on soils with high amounts of surface residues, with no-till management, or on wet soils. It is also useful for soils that test low-to-medium in phosphorus and potassium (or even higher). Banding fertilizer at planting, usually called **starter fertilizer,** may be a good idea even in warmer climates when planting early. It still might be cool enough to slow root growth and release of nutrients from organic matter. Including nitrogen in the band appears to help roots use fertilizer phosphorus more efficiently. Starter fertilizer for very low fertility soils frequently contains other nutrients, such as sulfur, zinc, boron, or manganese.

Splitting nitrogen applications is a good management practice — especially on sandy soils, where nitrate is easily lost by leaching, or on heavy loams and clays, where it is easily lost by denitrification. Some nitrogen is applied before planting or in the band as starter fertilizer, and the rest is applied as a **sidedress** or topdress during the growing season. Sometimes, split applications of potassium are recommended for very sandy soils with low organic matter, especially if there has been enough rainfall to cause potassium to leach into the subsoil. Unfortunately, relying on sidedressing nitrogen can increase risk of reduced yields if the weather is too wet to apply the fertilizer (and you haven't put on enough preplant or as starter) or too dry following an application. Then the fertilizer stays on the surface instead of washing into the root zone.

Once the soil nutrient status is optimal, try to balance farm nutrient inflows and outflows. When nutrient levels, especially phosphorus, are in the high or very high range, stop application and try to "draw down" soil test levels.

Tillage and Fertility Management: To Incorporate or Not?

With systems that provide some tillage, such as moldboard plow and harrow, disk harrow alone, chisel plow, zone-till, and ridge-till, it is possible to incorporate fertilizers and amendments. However, when using no-till production systems, it is not possible to mix materials into the soil to uniformly raise the fertility level in that portion of the soil where roots are especially active.

Soil Tests

Soil tests, one of the key nutrient management tools, are discussed in detail in chapter 19.

The advantages of incorporating fertilizers and amendments are numerous. Significant quantities of ammonia may be lost by volatilization when the most commonly used solid nitrogen fertilizer, urea, is left on the soil surface. Also, nutrients remaining on the surface after application are much more likely to be lost

---

Oxide vs. Elemental Forms?

When talking about using fertilizer phosphate or potash, the oxide form is usually assumed. This is used in all recommendations and when you buy fertilizer. When you put 100 lbs. of potash per acre, you actually applied 100 lbs. of K_2O — that's the equivalent of 83 lbs. of elemental potassium. Of course, you're not really using K_2O but rather something like muriate of potash (KCl). It's the same idea for phosphate — and 100 lbs. of P_2O_5 per acre is the same as 44 lbs. of P — and you're really using fertilizers like concentrated superphosphate (that contains a form of calcium phosphate) or ammonium phosphate.

---

in runoff during rain events. Although the amount of runoff is usually lower with reduced tillage systems than with conventional tillage, the concentration of nutrients in the runoff may be quite a bit higher.

If you are thinking about changing from conventional tillage to no-till or other forms of reduced tillage, you might consider incorporating needed lime, phosphate, and potash, as well as manures and other organic residues, before making the switch. It's the last chance to easily change the fertility of the top 8 or 9 inches of soil.

Sources

Ontario Ministry of Agriculture, Food, and Rural Affairs (OMAFRA). 1997. *Nutrient Management*. Best Management Practices Series. Available from the Ontario Federation of Agriculture, Toronto, Ontario (Canada).

Parnes, R. 1990. *Fertile Soil: A Grower's Guide to Organic and Inorganic Fertilizers*. agAccess, Davis, CA.

17

Management of Nitrogen and Phosphorus

*...an economical use of fertilizers requires that they
merely supplement the natural supply in the soil,
and that the latter should furnish the larger part of
the soil material used by the crop.*

—T.L. Lyon and E.O. Fippin, 1909

The management of nitrogen and phosphorus is discussed together because both are needed in large amounts by plants and both can cause environmental harm when present in excess. We don't want to do a good job of managing one and, at the same time, do a poor job with the other. The main environmental concern with nitrogen is the leaching of soil nitrate to groundwater and excess nitrogen in runoff. The drinking of high-nitrate groundwater is a health hazard to infants and young animals. In addition, nitrate stimulates the growth of algae and aquatic plants just like it does for agricultural plants. The growth of plants in many brackish estuaries and saltwater environments is believed to be limited by nitrogen. So, when nitrate leaches through soil, or runs off the surface and is discharged into streams, eventually reaching water bodies like the Gulf of Mexico or the Chesapeake Bay, undesirable microorganisms flourish. In addition, the algal blooms that result from excess nitrogen and phosphorus cloud water, blocking sunlight to important underwater grasses that are home to numerous species of young fish, crabs, and other bottom-dwellers.

Phosphorus is the nutrient that appears to limit the growth of freshwater aquatic weeds and algae. Phosphorus damages the environment when excess amounts are added to a lake from human activities (agriculture, rural home septic tanks, or urban sewage or runoff). This increases algae growth, making fishing, swimming, and boating unpleasant or difficult. When excess aquatic organisms die, decomposition removes oxygen from water and leads to fish kills.

All farms should work to have the best nitrogen and phosphorus management possible —

for economic as well as environmental reasons. This is especially important near bodies of water that are susceptible to accelerated weed or algae growth (eutrophication). However, don't forget that nutrients from farms in the Midwest are contributing to problems in the Gulf of Mexico — over 1,000 miles away.

There are major differences between the way nitrogen and phosphorus behave in soils (see table 17.1 and figure 17.1 below). Besides fertilizer sources, nitrogen is readily available to plants only from decomposing organic matter, while plants get their phosphorus from both organic matter and soil minerals. Nitrate, the primary form in which plants use nitrogen, is very mobile in soils; phosphorus movement in soil is very limited.

Most unintentional nitrogen loss from soils occurs when nitrate leaches or is converted into gases during denitrification, or when surface ammonium is volatilized. Large amounts of nitrate may leach from sandy soils, while denitrification is generally more important in heavy loams and clays. On the other hand, almost all unintended phosphorus loss from soils is carried away in sediments eroded from fields (see figure 17.1 for a comparison between relative pathways for nitrogen and phosphorus losses). Except when coming from highly manured fields, phosphorus losses from healthy grasslands are usually quite low — mainly as dissolved phosphorus in the runoff waters — because both runoff water and sediment loss are very low. Biological nitrogen fixation carried on in the roots of legumes and by some free-living bacteria actually adds new nitrogen to soil, but there is no equivalent reaction for phosphorus or any other nutrient.

Improving nitrogen and phosphorus management can help reduce reliance on commercial fertilizers. A more balanced system with good rotations and more active organic matter should provide a large proportion of crop nitrogen and phosphorus needs. Better soil structure and attention to use of appropriate cover crops can lessen loss of nitrogen and phosphorus by re-

TABLE 17.1
Comparing Soil Nitrogen and Phosphorus

NITROGEN		PHOSPHORUS
Nitrogen becomes available from decomposing soil organic matter.	vs.	Phosphorus becomes available from decomposing soil organic matter and from mineral forms.
Mostly available to plants as nitrate (NO_3^-) — a form that is very mobile in soils.	vs.	Available mainly as dissolved phosphate in soil water — but little present in solution even in fertile soils and is not mobile.
Nitrate can be easily lost by leaching to groundwater or conversion to gases (N_2, N_2O).	vs.	Phosphorus is mainly lost from soils by runoff and erosion.
Can add nitrogen to soils by biological nitrogen fixation (legumes).	vs.	No equivalent reaction can add new phosphorus to soil, although many bacteria and some fungi help make phosphorus more available.

ducing leaching, denitrification, and/or runoff. Reducing the loss of these nutrients is an economic benefit to the farm and, at the same time, an environmental benefit to society. The greater nitrogen availability may be thought of as a "fringe benefit" of a farm with an ecologically based cropping system. In addition, the manu-

facture, transportation, and application of nitrogen fertilizers is very energy intensive. Of all the energy used to produce corn (including the manufacture and operation of field equipment), the manufacture and application of nitrogen fertilizer represents close to 40 percent. So relying more on biological fixation of nitrogen reduces depletion of a non-renewable resource. Although energy has been relatively inexpensive for many years, it may very well become more expensive in the future. Although phosphorus fertilizers are less energy consuming to produce, a reduction in their use helps preserve this non-renewable resource.

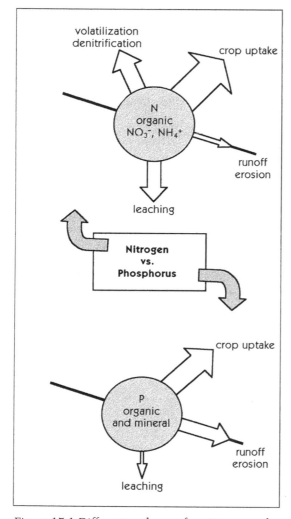

Figure 17.1 Different pathways for nitrogen and phosphorus losses from soils (relative amounts indicated by width of arrows). Based on an unpublished diagram by D. Beegle.

MANAGEMENT OF NITROGEN AND PHOSPHORUS

Nitrogen and phosphorus behave very differently in soils, but many of the management strategies are actually the same or are very similar. Management strategies include the following:

a) Take all nutrient sources into account.
- Use soil tests to assess available nutrients.
- Use manure tests to determine nutrient contributions.
- Consider nutrients in decomposing crop residues (for N only).

b) Reduce losses / enhance uptake.
- Use nutrient sources more efficiently.
- Use localized placement of fertilizers whenever possible.
- Split fertilizer application if leaching or denitrification losses are a problem (for N only).
- Apply nutrients when leaching or runoff threats are minimal.

- Reduce tillage.
- Use cover crops.
- Include perennial forage crops in rotation.

c) Balance farm imports and exports once crop needs are being met.

Taking All Nutrient Sources Into Account

Soil testing for nitrogen and phosphorus and interpreting soil test results are discussed in chapter 19.

Credit nutrients in manures, decomposing sods, and other organic residues. Before applying commercial fertilizers or other off-farm nutrient sources, you should properly credit the various on-farm sources of nutrients. In some cases, there is more than enough fertility in the on-farm sources to satisfy crop needs. If manure is applied before sampling soil, the contribution of much of the manure's phosphorus and all its potassium should be reflected in the soil test. One nitrogen soil test, the Pre-Sidedress Nitrate Test (PSNT), reflects the nitrogen contribution of the manure (see chapter 19 for a description of nitrogen soil tests). The only way to really know the nutrient value of a particular manure is to have it tested before applying it to the soil. Many soil test labs will also analyze manures for their fertilizer value. (Without testing the manure or the soil following application, estimates can be made based on average manure values, such as those given in table 9.1, p.79.) Because significant nitrogen losses can occur in as little as one or two days after manure application, the way to derive the full nitrogen benefit from manure is to incorporate it as soon as possible. Much of the manure-nitrogen made

available to the crop is in the ammonium form, and losses occur as ammonium is volatilized when manures dry on the soil surface. A significant amount of the manure's nitrogen also may be lost when application is a long time before crop uptake occurs. About half of the nitrogen value of a fall manure application — even if incorporated — may be lost by the time of greatest crop need the following year.

Legumes, as either part of rotations or as cover crops, and well-managed grass sod crops can add nitrogen to the soil for use by the following crops (table 17.2). Nitrogen fertilizer decisions should take into account the amount of nitrogen contributed by manures, decomposing sods, and cover crops. If you correctly filled out the form that accompanies your soil sample, the recommendation you receive may take these sources into account. However, soil testing labs may not take these into account when making

TABLE 17.2
Examples of Nitrogen Credits for Previous Crops

Previous Crop	N Credits (lbs./acre)
corn and most other crops	0
soybeans*	0 to 40*
grass (low level of management)	40
grass (intensively managed)	70
2-yr. stand red or white clover	70
3-yr. alfalfa stand (20–60% legume)	70
3-yr. alfalfa stand (>60% legume)	120
hairy vetch cover crop excellent growth	110

* Some labs give 30 or 40 lbs. N credit for soybeans, while others give no N credit.

their recommendations; most do not even ask whether you've used a cover crop. If you can't find help deciding how to credit nutrients in organic sources, take a look at chapters 9 (animal manures), 10 (rotations), and 11 (cover crops). For an example of crediting the nutrient value of manure and cover crop, see the section "Making Adjustments to Fertilizer Application Rates" in chapter 19 (p. 198).

Rely on legumes to supply nitrogen to following crops. Nitrogen is the only nutrient for which you can "grow" your own supply. High-yielding legume cover crops, such as hairy vetch and crimson clover, can supply most, if not all the nitrogen needed by the following crop. Growing a legume as a forage crop in rotation (alfalfa, alfalfa/grass, clover, clover/grass) also can provide much, if not all of the nitrogen for row crops. The nitrogen-related aspects of both cover crops and rotations with forages were discussed in previous chapters (chapters 10 and 11).

Animals on the farm or on nearby farms? If you have ruminant animals on your farm or on nearby farms for which you can grow forage crops, (and perhaps use the manure on your farm), there are many possibilities for actually eliminating the need to use nitrogen fertilizers. A forage legume, such as alfalfa, red clover, or white clover or a grass/legume mix, can supply substantial nitrogen for the following crop. Frequently, nutrients are imported onto animal farms

TABLE 17.3
Comparison of Nitrogen and Phosphorus Management Practices

NITROGEN	PHOSPHORUS
Use the PSNT or other reliable nitrogen soil test (and follow recommendations).	Soil test regularly (and follow recommendations).
Test manures and credit their nitrogen contribution.	Test manures and credit their phosphorus contribution.
Use legume forage crops in rotation and/or legume cover crops to fix nitrogen for following crops *and* properly credit legume nitrogen contribution to following crops.	No equivalent practice available.
Time N applications as close to crop uptake as possible.	Time P application to reduce runoff potential.
Reduce tillage in order to leave residues on the surface and decrease runoff and erosion.	Reduce tillage in order to leave residues on the surface and decrease runoff and erosion.
Sod-type forage crops in rotation reduce nitrate leaching and runoff.	Sod-type forage crops in rotation reduce the amount of runoff and erosion losses of phosphorus.
Grass cover crops, such as winter rye, capture soil nitrates left over following the economic crop.	Grass cover crops, such as winter rye, protect soil against erosion.
Make sure that excessive nitrogen is not coming onto the farm (biological nitrogen fixation + fertilizers + feeds).	After soil tests are in optimal range, balance farm phosphorus flow (don't import much more onto farm than is being exported).

as various feeds (usually grains and soybean meal mixes). This means that the manure from the animals will contain nutrients imported from outside the farm and this reduces the need to purchase fertilizers.

No animals? Although land constraints don't usually allow it, some vegetable farmers grow a forage legume for one or more years as part of a rotation, even when they are not planning to sell the crop or feed it to animals. They do so to rest the soil and to enhance the soil's physical properties and nutrient status. Also, some cover crops, such as hairy vetch — growing off-season in the fall and early spring — can provide sufficient nitrogen for some of the high-demanding summer annuals. It's also possible to undersow sweetclover and then plow it under the next July to prepare for fall brassica crops.

Reduce Nitrogen and Phosphorus Losses

Use nitrogen and phosphorus fertilizers more efficiently. If you've worked to build and maintain soil organic matter, you should have plenty of active organic materials present. These readily decomposable small fragments provide nitrogen and phosphorus as they are decomposed, reducing the amount of fertilizer that's needed.

The timing and method of application of commercial fertilizers and manures affect the efficiency of use by crops and the amount of loss from soils — especially in humid climates. In general, it is best to apply fertilizers close to the time when they are needed by plants. Losses of fertilizer and manure nutrients are also frequently reduced by soil incorporation with tillage.

If you're growing a crop for which a reliable in-season nitrogen test is available (see discussion in chapter 19) then you can hold off applying fertilizer until the nitrogen test indicates a need. At that point, apply nitrogen as a sidedress. Otherwise, you may need to broadcast some nitrogen before planting to supply sufficient nutrition until the soil test indicates if there is need for more nitrogen (applied as a sidedress). For row crops in colder climates, about 15 to 20 lbs. of starter N per acre (in a band at planting), is highly recommended.

Some of the nitrogen in surface-applied urea, the cheapest and most commonly used solid nitrogen fertilizer, is lost as a gas if it is not rapidly incorporated into the soil. If as little as $1/4$-inch rain falls within a few days of surface urea application, nitrogen losses are usually less than 10 percent. However, losses may be as large as 30 percent or more in some cases (50 percent loss may occur following surface application to a calcareous soil that is over pH 8). When urea is used for no-till systems, it can be placed below the surface. When fertilizer is broadcast as a topdress on grass or row crops, you might consider the economics of using ammonium nitrate. Although it is more costly than urea per unit of nitrogen, nitrogen in ammonium nitrate is generally not lost as a gas when left on the surface. Anhydrous ammonia, the least expensive source of nitrogen fertilizer, causes large changes in soil pH in and around the injection band. The pH increases for a period of weeks, many organisms are killed, and organic matter is rendered more soluble. Eventually, the pH decreases and the band is repopulated by soil organisms. However, significant nitrogen losses can occur when anhydrous is applied in a soil that is too dry or too wet. Even if stabilizers are used, anhydrous applied long before a crop needs it significantly increases the amounts of nitrogen that may be lost in humid regions.

If the soil is very deficient in phosphorus, phosphorus fertilizers are commonly incorporated into the soil to raise the general level of the nutrient. Incorporation is not possible with no-till systems and, if the soil was initially very deficient, some phosphorus fertilizer should have been incorporated before starting no-till. Nutrients accumulate near the surface of reduced tillage systems when fertilizers or manures are repeatedly surface-applied.

In soils with optimal phosphorus levels, some phosphorus fertilizer is still recommended, along with nitrogen application, for row crops in cool regions. (Potassium is also commonly recommended under these conditions.) Frequently, the soils are cold enough in the spring to slow down both root development and mineralization of phosphorus from organic matter, reducing phosphorus availability to seedlings. This is probably why it is a good idea to use some starter phosphorus in these regions — even if the soil is in the optimal phosphorus soil test range.

Use perennial forages (sod-forming crops) in rotations. As we've discussed a number of times, rotations that include a perennial forage crop help reduce the amount of runoff and erosion; build better soil tilth; break harmful weed, insect, and nematode cycles; and build soil organic matter. Decreasing the emphasis on row crops in a rotation and including perennial forages also helps decrease leaching losses of nitrate. This happens for two main reasons:

1) There is less water leaching under a sod because it uses more water over the entire growing season than does an annual row crop (which has a bare soil in the spring and after harvest in the fall).

2) Nitrate concentrations under sod rarely reach anywhere near those under row crops.

So, whether the rotation includes a grass, a legume, or a legume/grass mix, the amount of nitrate leaching to groundwater is usually reduced. (A critical step, however, is the conversion from sod to row crop. When a sod crop is plowed, a lot of nitrogen is mineralized. If this occurs many months before the row crop can use nitrogen, high nitrate leaching and denitrification losses occur.) Using grass, legume, or grass/legume forages in the rotation also helps with phosphorus management because of the reduction of runoff and erosion and the effects on soil structure for the following crop.

Use cover crops to prevent nutrient losses. High levels of soil nitrate may be left at the end of the growing season if drought causes a poor crop year or if excess nitrogen fertilizer or manure has been applied. The potential for nitrate leaching and runoff can be reduced greatly if you sow a fast-growing cover crop like winter rye. One option available when using cover crops to help manage nitrogen is to use a combination of a legume and grass. The combination of hairy vetch and winter rye works well in the mid-Atlantic region. When nitrate is scarce, the vetch does much better than the rye and a large amount of nitrogen is fixed for the next crop. On the other hand, the rye competes well with the vetch when nitrate is plentiful, and less nitrogen is fixed (of course, less is needed) and much of the nitrate is tied up in the rye and stored for future use.

In general, having any cover crop on the soil during the off season is helpful for phosphorus management. A cover crop that establishes quickly and helps protect the soil against erosion will help reduce phosphorus losses.

Reduce tillage. Because most phosphorus is lost from fields by erosion of sediments, environmentally sound phosphorus management should include reduced tillage systems. Leaving residues on the surface and maintaining stable soil aggregation and lots of large pores helps water to infiltrate into soils. When runoff does occur, less sediment is carried along with it than if conventional plow-harrow tillage is used. Reduced tillage, by decreasing runoff and erosion, usually decreases both phosphorus and nitrogen losses from fields.

Working Towards Balancing Nutrient Imports and Exports

Nitrogen and phosphorus are lost from soils, in many ways, including runoff that takes both nitrogen and phosphorus, leaching of nitrate (and sometimes significant amounts of phosphorus), denitrification, and volatilization of ammonia from surface applied urea and manures. Even if you take all precautions to reduce unnecessary losses, some loss of nitrogen and phosphorus will occur anyway. While you can easily overdo it with fertilizers, use of more nitrogen and phosphorus than needed also occurs on many livestock farms that import a significant proportion of their feeds. If a forage legume, such as alfalfa, is an important part of the rotation, the combination of biological nitrogen fixation plus imported nitrogen in feeds may exceed the farm's needs. A reasonable goal for farms with a large net inflow of nitrogen and phosphorus would

Reducing tillage usually leads to marked reductions in nitrogen and phosphorus loss in runoff and nitrate leaching loss to groundwater. However, there are two complicating factors that should be recognized:

✓ If intense storms occur soon after application of surface-applied urea or ammonium nitrate, nitrogen is more likely to be lost via leaching than if it had been incorporated. Much of the water will flow over the surface of no-till soils, picking up nitrate and urea, before entering wormholes and other channels. It then easily moves deep into the subsoil.

✓ Phosphorus accumulates on the surface of no-till soils (because there is no incorporation of broadcast fertilizers, manures, crop residues, or cover crops). Although there is less runoff and fewer sediments and less total phosphorus lost with no-till, the concentration of dissolved phosphorus in the runoff may actually be higher than for conventionally tilled soils.

be to try to reduce "imports" of these nutrients on farms (including legume nitrogen), or increase exports, to a point closer to balance.

On crop farms, as well as animal farms with low numbers of animals per acre, it's fairly easy to bring inflows and outflows into balance by properly crediting nitrogen from the previous crop and nitrogen and phosphorus in manure. On the other hand, it is a more challenging problem when there are a large number of animals for a given land base and a large percentage of the feed must be imported. This happens frequently on "factory"-type animal production facilities, but can also happen on family-sized farms. At some point, thought needs to be given

Managing High Phosphorus Soils

High-phosphorus soils occur because of a history of either excessive applications of phosphorus fertilizers or — more commonly — application of lots of manure. This is a problem on livestock farms with limited land and where a medium-to-high percentage of feed is imported. The nutrients imported in feeds may greatly exceed nutrients exported in animal products. In addition, where manures or composts are used at rates required to provide sufficient nitrogen to crops, more phosphorus than needed usually is added. It's probably a good idea to reduce the potential for phosphorus loss from all high-phosphorus soils. However, it is especially important to reduce the risk of environmental harm from those high-phosphorus soils that are also likely to produce significant runoff (because of steep slope, fine texture, poor structure, or poor drainage).

There are a number of practices that should be followed with high-phosphorus soils:

✓ First, deal with the "front end" and reduce animal phosphorus intake to the lowest levels needed. A recent survey found that the average dairy herd in the U.S. is fed about 25 percent more phosphorus than recommended by the standard authority (the National Research Council, NRC). In addition, research indicates that the NRC recommendations may be 10 to 15 percent higher than actually needed. This is costing dairy farmers about $3,500 to feed a 100-cow herd supplemental phosphorus that the animals don't need and that only ends up as a potential pollutant!

✓ Second, reduce or eliminate applying extra phosphorus. For a livestock farm, this may mean obtaining the use of more land to grow crops and to spread manure over a larger land area. For a crop farm, this may mean using legume cover crops and forages in rotations to supply nitrogen without adding phosphorus. The cover crops and forage rotation crops are also helpful to build up and maintain good organic matter levels in the absence of importing manures or composts or other organic material from off the farm. The lack of imported organic sources of nutrients (to try to reduce phosphorus imports) means that a crop farmer will need more creative use of crop residues, rotations, and cover crops to maintain good organic matter levels.

✓ Third, reduce runoff and erosion to minimal levels. Phosphorus usually is only a problem if it gets into surface waters. Anything that helps water infiltration or impedes water and sediments from leaving the field decreases problems caused by high-phosphorus soils — reduced tillage, strip cropping along the contour, cover crops, grassed waterways, riparian buffer strips, etc. [Note: Significant phosphorus losses in tile drainage water have been observed, especially from fields where large amounts of liquid manure are applied.]

✓ Fourth, continue to monitor soil phosphorus levels. Soil test phosphorus will slowly decrease over the years, once phosphorus imports as fertilizers, organic amendments, or feeds are reduced or eliminated. Testing soils every two or three years should be done for other reasons anyway. So, just remember to keep track of soil test phosphorus to confirm that levels are decreasing.

to either expanding the farm's land base or exporting some of the manure to other farms. Another option is to compost the manure — which makes it easier to transport or sell and causes some nitrogen losses during the composting process — stabilizing the remaining nitrogen before application. On the other hand, the availability of phosphorus in manure is not greatly affected by composting. That's why using compost to supply a particular amount of "available" nitrogen usually results in applications of larger total amounts of phosphorus than plants need.

Phosphorus and Potassium Can Get too High When Using Organic Sources

Manures and other organic amendments are frequently applied to soils at rates estimated to satisfy nitrogen needs of crops. This commonly adds more phosphorus and potassium than the crop needs. After many years of continuous application of these sources to meet nitrogen needs, soil test levels for phosphorus and potassium may be in the very high (excessive) range. Although there are a number of ways to deal with this issue, all solutions require reduced applications of fertilizer phosphorus and phosphorus-containing organic amendments. If it's a farm-wide problem, some manure may need to be exported and nitrogen fertilizer or legumes relied on to provide nitrogen to grain crops. Sometimes, it's just a question of better distribution of manure around the various fields — getting to those fields far from the barn more regularly. Changing the rotation to include crops, such as alfalfa, for which no manure nitrogen is needed can help. However, if you're raising livestock on a limited land base, you should make arrangements to remove some of the manure from the farm.

SOURCES

Brady, N.C., and R.R. Weil. 1999. *The Nature and Properties of Soils.* 12th ed. Macmillan Publ. Co. New York, NY.

Jokela, B., F. Magdoff, R. Bartlett, S. Bosworth, and D. Ross. 1998. *Nutrient Recommendations for Field Crops in Vermont.* UVM Extension, University of Vermont. Burlington, VT.

Magdoff, F.R. 1991. Understanding the Magdoff pre-sidedress nitrate soil test for corn. *Journal of Production Agriculture* 4:297–305.

National Research Council. 1988. Nutrient requirements of dairy cattle, 6th rev. ed., National Academy Press, Washington, D.C.

Sharpley, A.N. 1996. Myths about phosphorus. NRAES-96. *Proceedings from the Animal Agriculture and the Environment North American Conference, Dec. 11–13, Rochester, NY.* Northeast Region Agricultural Engineering Service. Ithaca, NY.

Vigil, M.F., and D.E. Kissel. 1991. Equations for estimating the amount of nitrogen mineralized from crop residues. *Soil Science Society of America Journal* 55:757–761.

18

Other Fertility Issues: Nutrients, CEC, Acidity and Alkalinity

The potential available nutrients in a soil,
whether natural or added in manures or fertilizer,
are only in part utilized by plants...

—T.L. LYON AND E.O. FIPPIN, 1909

OTHER NUTRIENTS

Although farmers understandably emphasize nitrogen and phosphorus — because of the large quantities used and the potential for environmental problems — additional nutrient and soil chemical issues remain important. In most cases, the overuse of these other fertilizers and amendments doesn't cause problems for the environment, but inappropriate use may waste money and reduce yields. There are also animal health considerations. For example, excess potassium in feeds for dry cows (between lactations) results in metabolic problems, and low magnesium availability to dairy or beef cows in early lactation can cause grass tetany. As with most other issues we have discussed, focusing on the management practices that build up and maintain soil organic matter will help eliminate many problems or, at least, make them easier to manage.

Potassium is one of the N-P-K "big three" primary nutrients needed in large amounts and frequently not present in sufficient quantities for plants. Potassium availability to plants is sometimes decreased when liming a soil to increase the pH by one or two units. The extra calcium, as well as the "pull" on potassium exerted by the new cation exchange sites (see discussion of CEC), contribute to lower potassium availability. Problems with low potassium levels are usually dealt with easily by applying muriate of potash (potassium chloride), potassium sulfate, or Sul-Po-Mag or K-Mag (potassium and magnesium sulfate). Manures also usually contain large quantities of potassium.

Magnesium deficiency is easily corrected if the soil is acidic by using a high-magnesium (dolomitic) limestone to raise soil pH (see discussion of soil acidity below). Otherwise, Sul-Po-Mag is one of the best choices for correcting such a deficiency.

Calcium deficiencies are usually associated with low pH soils and soils with low CECs. The best remedy is usually to lime and build up the soil's organic matter. However, some important crops, such as peanuts, potatoes and apples, commonly need added calcium. Calcium additions also may be needed to help alleviate soil structure and nutrition problems of sodic soils (see below). In general, if the soil does not have too much sodium, is properly limed and has a reasonable amount of organic matter, there will be no advantage to adding a calcium source, such as gypsum. However, soils with very low aggregate stability may sometimes benefit from the extra salt concentration associated with surface gypsum applications. This is not a calcium nutrition effect, but a stabilizing effect of the dissolving gypsum salt. Higher soil organic matter and surface residues should do as well as gypsum to alleviate this problem.

Sulfur deficiencies are common on soils with low organic matter. Some soil testing labs around the country offer a sulfur soil test. (Those of you who grow garlic should know that a good supply of sulfur is important for the full development of garlic's pungent flavor — so garlic growers want to make sure there's plenty available to the crop.) Much of the sulfur in soils occurs as organic matter, so building up and maintaining good amounts of organic matter should result in sufficient sulfur nutrition for plants. Although reports of crop response to added sulfur in the Northeast are rare, it is thought that deficiencies of this element may become more common

now that there is less sulfur air pollution originating in the Midwest. Some fertilizers used for other purposes, such as Sul-Po-Mag and ammonium sulfate, contain sulfur. Calcium sulfate (gypsum) also can be applied to remedy low soil sulfur. The amounts used on sulfur-deficient soils are typically 20 to 25 lbs. sulfur/acre.

Zinc deficiencies occur with certain crops on soils low in organic matter and in sandy soils or those with a pH near neutral. Zinc problems are sometimes noted on silage corn when manure hasn't been applied for a while. It also can be deficient following topsoil removal from parts of fields as land is leveled for furrow irrigation. Sometimes, crops outgrow the problem as the soil warms up and organic sources become more available to plants. Zinc sulfate (about 35 percent zinc) applied to soils is one of the materials used to correct zinc deficiencies. If the deficiency is due to high pH, or if an orchard crop is zinc-deficient, a foliar application is commonly used. If a soil test before planting an orchard reveals low zinc levels, zinc sulfate should be soil-applied.

Boron deficiencies show up in alfalfa when growing on eroded knolls where topsoil and organic matter have been lost. Root crops seem to need higher soil boron levels than many other crops. Cole crops, apples, celery and spinach are also sensitive to low boron levels. The most common fertilizer used to correct a boron deficiency is sodium tetraborate (about 15 percent boron). Borax (about 11 percent boron), a compound containing sodium borate, also can be used to correct boron deficiencies. On sandy soils low in organic matter, boron may be needed on a routine basis.

Manganese deficiency, usually associated with soybeans and cereals on high pH soils and vegetables grown on muck soils, is corrected with the use of manganese sulfate (about 27 percent manganese). About 10 lbs. of water-soluble manganese per acre should satisfy plant needs for a number of years. Up to 25 lbs. per acre of manganese is recommended, if the fertilizer is broadcast on a very deficient soil. Natural, as well as synthetic, chelates (at about 5 to 10 percent manganese) usually are applied as a foliar spray.

Iron deficiency occurs when blueberries are grown on moderate to high pH soils. Iron deficiency also sometimes occurs on soybeans, wheat, sorghum, and peanuts growing on high pH soils. Iron (ferrous) sulfate or chelated iron are used to correct iron deficiency. Both manganese and iron deficiencies are frequently corrected by using foliar application of inorganic salts.

CATION EXCHANGE CAPACITY MANAGEMENT

The CEC in soils is due to well humified ("very dead") organic matter and clay minerals. The total CEC in a soil is the sum of the CEC due to organic matter and CEC due to the clays. In fine-textured soils with medium to high CEC-type clays, much of the CEC may be due to clays. On the other hand, in sandy loams with little clay, or in some of the soils of the southeastern U.S. containing clays with low CEC, organic matter may account for the overwhelming fraction of the total CEC.

There are two practical ways to increase the ability of soils to hold nutrient cations, such as potassium, calcium, magnesium, and ammonium:

- Add organic matter by many of the methods discussed earlier in Part Two.
- If the soil is too acidic, use lime (see below) to raise its pH to the high end of the range needed for the crops you grow.

One of the benefits of liming acid soils is to increase soil CEC! Here's why: As the pH increases, so does the CEC of organic matter. As hydrogen (H^+) on humus is neutralized by liming, the site where it was attached now has a negative charge and can hold onto Ca^{++}, Mg^{++}, K^+, etc.

Estimating Organic Matter's Contribution to a Soil's CEC

The CEC of a soil is usually expressed in terms of the number of milliequivalents (me) of negative charge per 100 grams of soil. (The actual number of charges represented by one me is about 6 followed by 20 zeros.) A useful rule of thumb for estimating the CEC due to organic matter is as follows: for every pH unit above pH 4.5, there is 1 me of CEC in 100 gm of soil for every percent organic matter. (Don't forget that there will also be CEC due to clays.)

Example 1: pH = 5.0 and 3% SOM → (5.0 — 4.5) x 3 = 1.5 me/100g

Example 2: pH = 6.0 and 3% SOM → (6.0 — 4.5) x 3 = 4.5 me/100g

Example 3: pH = 7.0 and 3% SOM → (7.0 — 4.5) x 3 = 7.5 me/100g

Example 4: pH = 7.0 and 4% SOM → (7.0 — 4.5) x 4 = 10.0 me/100g

---

Many soil testing labs also will run CEC, if asked. However, there are a number of possible ways to do the test. Some labs determine what the CEC would be if the soil's pH was 7 or higher. They do this by adding the acidity that would be neutralized if the soil was limed to the current soil CEC. This is the CEC the soil *would* have at the higher pH, but is not the soil's current CEC. For this reason, some labs total the major cations actually held on the CEC (Ca^{++} + K^+ + Mg^{++}) and call it effective CEC. It is more useful to know the effective CEC — the actual current CEC of the soil — than CEC determined at a higher pH.

SOIL ACIDITY

Background

Plants have evolved under specific environments, which in turn influence their needs as agricultural crops. For example, alfalfa originated in a semiarid region where soil pH was high; alfalfa requires a pH in the range of 6.5 to 6.8 or higher (see figure 18.1 for common soil pH levels). On the other hand, blueberries, which evolved under acidic conditions, require a low pH to provide needed iron (iron is more soluble at low pH). Other crops, such as peanuts, watermelons, and sweet potatoes, do best in moderately acid soils in the range of pH 5 to 6. Most other agricultural plants do best in the range of pH 6 to 7.5.

Several problems may cause poor growth of acid-sensitive plants in low pH soils. The following are three common ones:

- aluminum and manganese are more soluble and can be toxic to plants;
- lack of calcium, magnesium, potassium, phosphorus or molybdenum (especially needed for nitrogen fixation by legumes); and
- a slowed decomposition of soil organic matter and decreased mineralization of nitrogen.

The problems caused by soil acidity are usually less severe, and the optimum pH is lower, if the soil is well supplied with organic matter. Organic matter helps to make aluminum less toxic and, of course, humus increases the soil's CEC. Soil pH will not change as rapidly in soils that are high in organic matter. Soil acidification is a natural process that is accelerated by acids produced in soil by most nitrogen fertilizers. Soil organic matter slows down acidification and buffers the soil's pH because it holds the acid hydrogen tightly. Therefore, more acid is needed to decrease the pH by a given amount

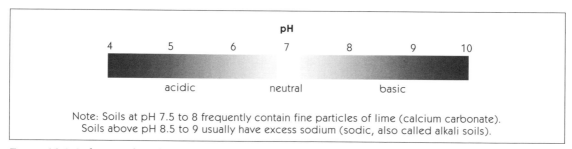

Note: Soils at pH 7.5 to 8 frequently contain fine particles of lime (calcium carbonate). Soils above pH 8.5 to 9 usually have excess sodium (sodic, also called alkali soils).

Figure 18.1 Soil pH and acid/base status.

BUILDING SOILS FOR BETTER CROPS

when a lot of organic matter is present. Of course, the reverse is also true — more lime is needed to raise the pH of high-organic matter soils by a given amount (see "Soil Acidity" box below).

Soil Acidity

BACKGROUND

✓ pH 7 is neutral.

✓ Soil with pH levels above 7 are alkaline, those less than 7 are acidic.

✓ The lower the pH, the more acidic is the soil.

✓ Soils in humid regions tend to be acidic, those in semi-arid and arid regions tend to be around neutral or are alkaline.

✓ Acidification is a natural process.

✓ Most commercial nitrogen fertilizers are acid forming, but many manures are not.

✓ Crops have different pH needs — probably related to nutrient availability or susceptibility to aluminum toxicity at low pH.

✓ Organic acids on humus and aluminum on the CEC account for most of the acid in soils.

MANAGEMENT

✓ Use limestone to raise soil pH (if magnesium is also low, use a high magnesium — or dolomitic — lime).

✓ Mix lime thoroughly into the plow layer.

✓ Spread lime well in advance of sensitive crops, if at all possible.

✓ If the lime requirement is high — some labs say greater than 2 tons and others say greater than 4 tons — consider splitting the application over two years.

✓ Reducing soil pH (making soil more acid) for acid-loving crops is done best with elemental sulfur (S).

Limestone application helps create a more hospitable soil for acid-sensitive plants in many ways, such as:

- neutralizing acids;
- adding calcium in large quantities (because limestone is calcium carbonate, $CaCO_3$);
- adding magnesium in large quantities if dolomitic limestone is used (containing carbonates of both calcium and magnesium);
- making molybdenum and phosphorus more available;
- helping to maintain added phosphorus in an available form;
- enhancing bacterial activity; and
- making aluminum and manganese less soluble.

Almost all the acid in acidic soils is held in reserve on the solids, with an extremely small amount active in the soil water. If all that we needed to neutralize was the acid in the soil water, a few handfuls of lime per acre would be enough to do the job, even in a very acid soil. However, tons of lime per acre are needed to raise the pH. The explanation for this is that almost all of the acid that must be neutralized in soils is reserve acidity associated with either organic matter or aluminum.

pH Management

Increasing the pH of acidic soils is usually accomplished by adding ground or crushed limestone. Three pieces of information are used to determine the amount of lime that's needed:

- What is the soil pH? Knowing this and the needs of the crops you are growing tell whether lime is needed and what target pH you are shooting for. If the soil pH is much lower than the pH needs of the crop, you need to use lime. But the pH value doesn't tell you how much lime is needed.

- What is the lime requirement needed to change the pH to the desired level? There are a number of different tests used by soil testing laboratories that estimate soil lime requirements. Most give the results in terms of tons/acre of agricultural grade limestone to reach the desired pH.
- Is the limestone you use very different from the one assumed in the soil test report? The fineness and the amount of carbonate present govern the effectiveness of limestone — how much it will raise the soil's pH. If the lime you will be using has an effective calcium carbonate equivalent that's very different from the one used as the base in the report, the amount applied may need to be adjusted upward (if the lime is very coarse or has a high level of impurities) or downward (if the lime is very fine and is high in magnesium and contains few impurities).

Testing labs usually use the information you provide about your cropping intentions and integrate the three issues (see left column) when recommending limestone application rates. There are laws governing the quality of limestone sold in each state. Soil testing labs give recommendations assuming the use of ground limestone that meets the minimum state standard.

Soils with more clay and more organic matter need more lime to change their pH (see figure 18.2). Although organic matter buffers the soil against pH decreases, it also buffers against pH increases when you are trying to raise the pH with limestone. Most states recommend a soil pH around 6.8 only for the most sensitive crops, such as alfalfa, and about pH 6.2 to 6.5 for many of the clovers. As pointed out above, most of the commonly grown crops do well in the range of pH 6.0 to 7.5.

There are other liming materials in addition to limestone. One of the commonly used ones in some parts of the U.S. is wood ash. Ash from a modern air-tight wood burning stove may have a fairly high calcium carbonate content (80 percent or higher). However, ash that is mainly black — indicating incompletely burned wood — may have as little as 40 percent effective calcium carbonate equivalent. Lime-sludge from wastewater treatment plants and fly ash sources may be available in some locations. Normally, minor sources like the ones mentioned above are not locally available in sufficient quantities to put much of a dent in the lime needs of a region. Because they might carry unwanted contaminants to the farm, be sure that liming materials are thoroughly evaluated for metals and field-tested before you use any new byproduct liming sources.

"Overliming" Injury

Sometimes there are problems when soils are limed, especially if a very acidic soil has been quickly raised to high pH levels. Decreased crop growth because of "overliming" injury is usually associated with lowered availability of phosphorus, potassium, or boron, although zinc and copper deficiencies can be produced by liming acidic sandy soils. If there has been a long history of use of triazine herbicides, such as atrazine, liming may release these chemicals and kill sensitive crops.

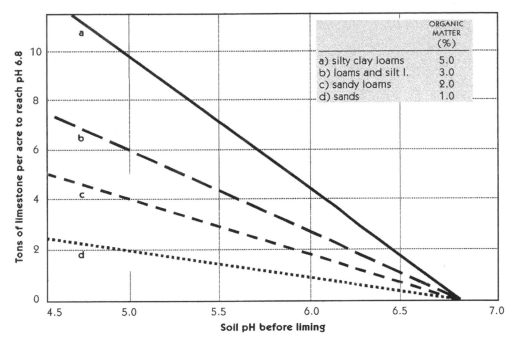

Figure 18.2 Examples of approximate lime needs to reach pH 6.8. Modified from Peech, 1961.

Need to Lower the Soil's pH?

When growing plants that require low pH, you may want to add acidity to the soil. This is probably only economically possible for blueberries and is most easily done with elemental sulfur (S), which is converted into an acid by soil microorganisms over a few months. For the examples in figure 18.2, the amounts of S needed to drop the pH <u>by one unit</u> would be approximately ¾ ton per acre for the silty clay loams, ½ ton per acre for the loams and silt loams, 600 lbs. per acre for the sandy loams, and 300 lbs. per acre for the sands. Sulfur should be applied the year before planting blueberries.

Alum (aluminum sulfate) may also be used to acidify soils. About six times more alum than elemental sulfur is needed to achieve the same pH change.

ARID REGION PROBLEMS: SODIC (ALKALI) AND SALINE SOILS

Special soil problems are found in arid and semi-arid regions, including soils that are high in salts, called saline soils, and those that have excessive sodium (Na⁺), called sodic soils. Sometimes these go together and the result is a saline-sodic soil. Saline soils usually have good soil tilth, but plants can't get the water they need because the high salt levels inhibit water-uptake. Sodic soils tend to have very poor physical structure because the high sodium levels cause clays to disperse, leading to the breaking apart of aggregates. As aggregates break down, these soils become difficult to work with and very unpleasant for plants. Wet sodic soils with an adequate amount of clay end up looking and behaving like chocolate pudding.

Saline and sodic soils are commonly found in the semi-arid and arid regions in the western U.S., with pockets of saline soils found near the coastline. Sometimes, the extra moisture accumulated during a fallow year in semi-arid regions causes field-seeps, which lead to the development of saline and sodic patches.

Before embarking on a program to improve saline or sodic soil, it is important to find out what is causing the problem. Do you have a high water table that contains salty water? Is there a saline or alkali "seep" in a portion of the field? Is it a generalized problem over the entire field without a high water table? If a high water table is causing salts to migrate upward to the root zone, installation of tile drainage may be necessary. Water surfacing in saline and alkali seeps usually can be reduced by changing from an alternate year fallow to a more intense annual cropping regime.

There are a number of ways to deal with saline soils that don't have shallow salty groundwater. One is to keep the soil continually moist. For example, by using drip irrigation with low salt water plus a surface mulch, the salt content will not get as high as it would if allowed to concentrate when the soil dries. Another is to grow crops or varieties of crops that are more tolerant of soil salinity. Saline-tolerant plants include bar-

ley, bermuda grass, oak, rosemary, and willow. However, the only way to get rid of the salt is to add sufficient water to wash it below the root zone. The amount of water needed to do this is related to the salt content of the irrigation water, expressed as electrical conductivity (ECw) and the salt content desired in the drainage water (ECdw). The amount of water needed can be calculated using the following equation:

$$\text{Water needed} = (\text{amount of water needed to saturate soil}) \times (ECw/ECdw)$$

The amount of extra irrigation water needed to leach salts is also related to the sensitivity of the plants that you're growing. For example, sensitive crops like onions and strawberries may have twice the leaching requirement as moder-

Salts are Present in All Soils

Salts of calcium, magnesium, potassium and other cations — along with the common negatively charged anions chloride, nitrate, sulfate, and phosphate — are found in all soils. However, in soils in subhumid and humid climates — with from an inch or two to well over 7 inches of water percolating beneath the root zone every year — salts don't usually accumulate to the levels where they can be harmful to plants. Even when high rates of fertilizers are used, salts usually only become a problem when you place large amounts in direct contact with seeds or growing plants. Salt problems frequently occur in greenhouse potting mixes because growers regularly irrigate their greenhouse plants with water containing fertilizers and may not add enough water to leach the accumulating salts out of the pot.

ately sensitive broccoli or tomatoes. Drip irrigation uses relatively low amounts of water, so lack of leaching may cause salt build-up even for moderately saline irrigation sources. This means that the leaching may need to occur during the growing season, but care is needed to prevent leaching of nitrate below the root zone.

For sodic soils, a calcium source is added — usually gypsum (calcium sulfate). The calcium replaces sodium held by the cation-exchange capacity. The soil is then irrigated so that the sodium can be leached deep in the soil. Because the calcium in gypsum easily replaces the sodium on the CEC, the amount of gypsum needed can be estimated as follows — for every milliequilivent of sodium that needs to be replaced to 1 foot, about 2 tons of agricultural grade gypsum is needed per acre. Adding gypsum to non-sodic soils doesn't help physical properties if the soil is properly limed, except for those soils containing easily dispersible clay that are also low in organic matter.

SOURCES

Hanson, B.R., S.R. Grattan, and A. Fulton. 1993. *Agricultural Salinity and Drainage.* Publication 3375, Div. of Agriculture and Natural Resources, University of California, Oakland, CA.

Magdoff, F.R. and R.J. Bartlett. 1985. Soil pH buffering revisited. *Soil Science Society of America Journal* 49:145–148.

Peech, M. 1961. *Lime Requirement vs. Soil pH Curves for Soils of New York State.* Mimeographed. Cornell University Agronomy Department, Ithaca, NY.

Pettygrove, G.S., S.R. Grattan, T.K. Hartz, L.E. Jackson, T.R. Lockhart, K.F. Schulbach, and R. Smith. 1998. *Production Guide: Nitrogen and Water Management for Coastal Cool-Season Vegetables.* Publication 21581, Division of Agriculture and Natural Resources, University of California, Oakland, CA.

Rehm, G. 1994. *Soil Cation Ratios for Crop Production.* North Central Regional Extension Publication 533. University of Minnesota Extension Service, St. Paul, MN.

Tisdale, S.L., W.I. Nelson, J.D. Beaton, and J.L. Havlin. 1993. *Soil Fertility and Fertilizers.* Macmillan Publishing Co., New York, NY.

19

Getting the Most
From Soil Tests

...the popular mind is still fixed on the idea that
a fertilizer is the panacea.

—J.L. HILLS, C.H. JONES, AND C. CUTLER, 1908

Although fertilizers and other amendments purchased from off the farm are not a panacea to cure all soil problems, they play an important role in maintaining soil productivity. Soil testing is the farmer's best means for determining which amendments or fertilizers are needed and how much should be used.

The soil test report provides the soil's nutrient and pH levels and, in arid climates, the salt and sodium levels. Recommendations for application of nutrients and amendments accompany most reports. They are based on soil nutrient levels, past cropping and manure management, and should be a customized recommendation based on the crop you plan to grow.

Soil tests — and proper interpretation of results — are a very important management tool for developing a farm nutrient management program. However, deciding how much fertilizer to apply — or the total amount of nutrients needed from various sources — is part science, part philosophy, and part art. Understanding soil tests and how to interpret them can help farmers better customize the test's recommendations. In this chapter, we'll go over sources of confusion about soil tests, discuss N and P soil tests, and then examine a number of soil tests to see how the information they provide can help you make decisions about fertilizer application.

TAKING SOIL SAMPLES

The usual time to take soil samples for general fertility evaluation is in the fall or in the spring, before the growing season has begun. These samples are analyzed for pH and lime requirement as well as phosphorus, potassium, and magnesium. Some labs also routinely analyze for selected micronutrients, such as boron, zinc, and manganese.

ACCURACY OF RECOMMENDATIONS BASED ON SOIL TESTS

Soil tests and their recommendations, although a critical component of fertility management, are not 100 percent accurate. Soil tests are an important tool, but need to be used by farmers and farm-advisors along with other information to make the best decision regarding amounts of fertilizers or amendments to apply.

Soil tests are an estimate of a limited number of plant nutrients, based on a small sample, which is supposed to represent many acres in a field. With soil testing, the answers aren't quite as certain as we might like them. A low potassium soil test indicates that you will *probably* increase yield by adding the nutrient. However, adding fertilizer may not increase crop yields in a field with a low soil test level. The higher yields may be prevented because the soil test is not calibrated for that particular soil (and the soil had sufficient potassium for the crop despite the low test level) or because of harm caused by poor drainage or compaction. Occasionally, using extra nutrients on a high-testing soil increases crop yields. Weather conditions may have made the nutrient less available than indicated by the soil test. So, it's important to use common sense when interpreting soil test results.

Guidelines for Taking Soil Samples

1. Don't wait until the last minute. The best time to sample for a general soil test is usually in the fall. Spring samples should be taken early enough to have results in time to properly plan nutrient management for the crop season.
2. Take cores from at least 15 to 20 spots randomly over the field to obtain a representative sample. One sample should not represent more than 10 to 20 acres.
3. Sample between rows. Avoid old fence rows, dead furrows, and other spots that are not representative of the whole field.
4. Take separate samples from problem areas, if they can be treated separately.
5. In cultivated fields, sample to plow depth.
6. Take two samples from no-till fields: one to a 6-inch depth for lime and fertilizer recommendations, and one to a 2-inch depth to monitor surface acidity.
7. Sample permanent pastures to a 3- to 4-inch depth.
8. Collect the samples in a clean container.
9. Mix the core samplings, remove roots and stones, and allow to air dry.
10. Fill the soil-test mailing container.
11. Complete the information sheet, giving all of the information requested. Remember, the recommendations are only as good as the information supplied.
12. Sample fields at least every three years. Annual soil tests will allow you to fine-tune nutrient management and may allow you to cut down on fertilizer use.

—Modified from The PennState Agronomy Guide, 1999.

BUILDING SOILS FOR BETTER CROPS

SOURCES OF CONFUSION
ABOUT SOIL TESTS

People may be easily confused about the details of soil tests, especially if they have seen results from more than one soil testing laboratory. There are a number of reasons for this, including:

- laboratories use a variety of procedures;
- labs report results differently; and
- different approaches are used to make recommendations based on soil test results.

Labs Use Varied Procedures

One of the complications with using soil tests to help determine nutrient needs is that testing labs across the country use a wide range of procedures. The main difference among labs is the solutions they use to extract the soil nutrients. Some use one solution for all nutrients, while others will use one solution to extract potassium, magnesium and calcium; another for P; and yet another for micronutrients. The various extracting solutions have different chemical compositions, so the amount of a particular nutrient that lab A extracts may be very different from the amount extracted by lab B. However, there are frequently good reasons to use a particular solution. For example, the Olsen test for phosphorus (see below) is more accurate for high-pH soils in arid and semi-arid regions than are the various acid-extracting solutions commonly used in more humid regions. Whatever procedure the lab uses, soil test levels must be calibrated with crop yield response to added nutrients. For example, do the yields really increase when you add phosphorus to a soil that tests low in P? In general, university or state labs in a given region use the same or similar procedures that have been calibrated for local soils and climate.

Labs Report Soil
Test Levels Differently

Different labs may report their results in different ways. Some use part per million (10,000 ppm = 1 percent); others use lbs./acre (they do this usually by using part per *two* million, which is twice the part per million level); and others use an index (for example, all nutrients are expressed on a scale of 1 to 100). In addition, some labs report phosphorus and potassium in the elemental form, while others use the oxide forms, P_2O_5 and K_2O.

Soil tests are not perfect.
All they indicate is whether or not
adding a nutrient is likely to result in
a yield increase of the crop
growing on that particular soil.

Most testing labs report results as both a number and a category, such as low, medium, optimum, high, very high. However, although most labs consider high to be above the amount needed (the amount needed is called optimum), some labs use *optimum* and *high* interchangeably. If the significance of the various categories is not clear on your report, be sure to ask. Labs should be able to furnish you with the probability of getting a response to added fertilizer for each soil test category.

Different Recommendation Systems

Even when labs use the same procedures, as is the case in most of the Midwest, different approaches to making recommendations lead to different amounts of recommended fertilizer. Three different philosophies are used to make fertilizer recommendations based on soil tests. One approach — the *sufficiency level system* — suggests there is a point, the sufficiency or critical soil test value, above which there is little likelihood of response to an added nutrient. Its goal is not to produce the highest yield every year, but, rather, to produce the highest average return over time from using fertilizers. Experiments that relate yield increases with added fertilizer to soil test level provide much of the evidence supporting this approach. As the soil test level increases from optimum to high, yields without adding fertilizer are closer to the maximum obtained by adding fertilizer (figure 19.1). Of course, farmers should be shooting for the maximum *economic* yields, which are slightly below the highest possible yields.

Another approach used by soil test labs — the *build-up and maintenance system* — calls for building up soils to high levels of fertility and then keeping them there by applying enough fertilizer to replace nutrients removed in harvested crops. It is used mainly for phosphorus, potassium, and magnesium recommendations.

The *basic cation saturation ratio system,* a method of estimating calcium, magnesium, and potassium needs, is based on the belief that crops yield best when calcium (Ca^{++}), magnesium (Mg^{++}), and potassium (K^+) — usually the dominant cations on the CEC — are in a particular balance. Although there are different versions of this system, most call for calcium to occupy about 60 to 80 percent of the CEC, whereas magnesium should be from 10 to 20 percent and potassium from 2 to 5 percent of the CEC. Great care is needed when using the base cation saturation ratio system. For example, the ratios

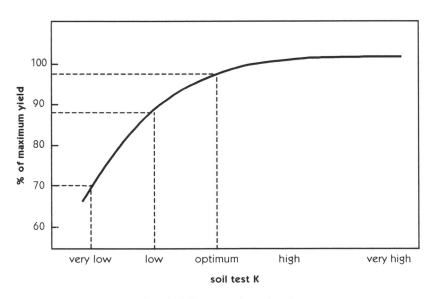

Figure 19.1 Percent of maximum yield with different soil test levels.

of the nutrients can be within the guidelines, but there may be such a low CEC (such as with a sandy soil that is very low in organic matter), that the amounts present are insufficient for crops. In addition, when there is a high CEC, there may be plenty of the nutrient, but the cation ratio system will call for adding more. This can be a problem with soils that are naturally high in magnesium, because the recommendations may call for high amounts of calcium and potassium to be added when none are really needed.

Research indicates that plants do well over a broad range of cation ratios, as long as there are sufficient supplies of potassium, calcium, and magnesium. However, there are occasions when the calcium-magnesium-potassium ratios are very out of balance. For example, when magnesium occupies more than 50 percent of the CEC in soils with low aggregate stability, using calcium sulfate may help restore aggregation. As mentioned previously, liming very acidic soils sometimes results in decreased potassium avail-ability and this would be apparent when using the cation ratio system. The *sufficiency system* would also call for adding potassium because of the low potassium levels in these very acid soils.

The *sufficiency level* approach is used by most fertility recommendation systems for potassium, magnesium, and calcium. It generally calls for lower application rates and is more consistent with the scientific data than the cation ratio system. The cation ratio system can be used successfully, if interpreted with care and common sense — not ignoring the total amounts present.

Labs sometimes use a combination of these systems, something like a hybrid approach. Some laboratories that use the sufficiency system will have a target for magnesium, but then suggest adding more if the potassium level is high. Others suggest that higher potassium levels are needed as the soil CEC increases. These are really hybrids of the sufficiency and cation ratio systems. At least one lab uses the sufficiency system for potassium and a cation ratio system for calcium and magnesium. Also, some labs assume that soils will *not* be tested annually. The recommendation that they give is, therefore, a combination of the sufficiency system (what is needed for this crop) with a certain amount added for maintenance. This is done to be sure there is enough fertility in the following year.

To estimate the percentages of the various cations on the CEC, the amounts need to be expressed in terms of quantity of charge. Some labs give both concentration by weight (ppm) and by charge (me/100g). If you want to convert from ppm to milliequivalent per 100 grams (me/100g), you can do it as follows:

$$\text{(Ca in ppm)}/200 = \text{Ca in me/100g}$$
$$\text{(Mg in ppm)}/120 = \text{Mg in me/100g}$$
$$\text{(K in ppm)}/390 = \text{K in me/100g}$$

As discussed previously (chapter 18), adding up the amount of charge due to calcium, magnesium, and potassium gives a very good estimate of the CEC for most soils for most soils above pH 5.5.

Crop Value, Fertilizer Rates, and Recommendation System

The value of your crop can have a major impact on the economics of over-applying fertilizer. As a general rule, the lower the per acre value of your crop, the greater the economic penalty for applying extra fertilizer (see box on p. 183). Farmers growing agronomic crops should take special care not to over-apply fertilizer.

As the soil test level of a particular nutrient increases, there is less chance that adding the nutrient will result in a greater yield. However, it may be worth adding fertilizer to high-value crops grown on soils with the same test levels that call for no fertilizer use low-value crops (figure 19.2). This difference should be reflected in the recommendations provided by soil testing laboratories.

Plant Tissue Tests

Soil tests are the most common means of assessing fertility needs of crops, but plant tissue tests are especially useful for nutrient management of perennial crops, such as apple, citrus and peach orchards, and vineyards. For most annuals, including agronomic and vegetable crops, tissue testing is not widely used, but can help diagnose problems. The small sampling window available for most annuals and an in-

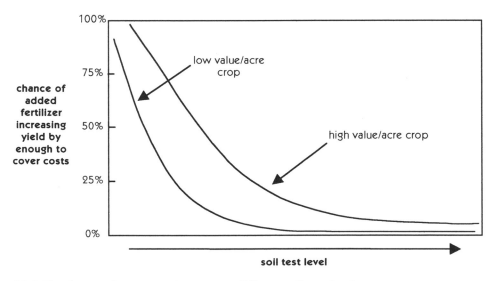

Figure 19.2 The chance of an economic return at different soil test levels.

Crop Value and Care with Fertilizer Rates

Most agronomic crops grown on large acreages are worth around $200 to $400 per acre and the fertilizer used may represent 30 to 40 percent of out-of-pocket growing costs. So, if you use 100 lbs. of N you don't need, that's about $30/acre and represents a major economic loss. Some years ago, one of the authors worked with two brothers who operated a dairy farm in northern Vermont that had high soil test levels of N, P, and K. Despite his recommendation that no fertilizer was needed, the normal practice was followed and $70 per acre worth of fertilizer N, P, and K was applied to their 200 acres of corn. The yields on 40 feet wide, no-fertilizer strips that they left in each field were the same as where fertilizer had been applied, so the $14,000 for fertilizer was wasted.

When growing fruit or vegetable crops — worth thousands of dollars per acre — fertilizers represent about 1 percent of the value of the crop and 2 percent of the costs. But when growing specialty crops (medicinal herbs, certain organic vegetables for direct marketing) worth over $10,000 per acre, the cost of fertilizer is dwarfed by other costs, such as hand labor. A waste of $30/acre in unneeded nutrients for these crops would cause a minimal economic penalty — assuming you maintain a reasonable balance between nutrients — but there may be environmental reasons against applying too much fertilizer. However, there may be a justification for using the build-up and maintenance approach for phosphorus and potassium on high-value crops because: a) the extra costs are such a small percent of total costs and b) there may occasionally be a higher yield because of this approach that would more than cover the extra expense of the fertilizer.

ability to effectively fertilize them once they are well established, except for N during early growth stages, limits the usefulness of tissue analysis for annual crops. However, leaf petiole nitrate tests are sometimes done on potato and sugar beets to help fine-tune in-season N fertilization. Petiole nitrate is also helpful for N management of cotton and for help managing irrigated vegetables, especially during the transition from vegetative to reproductive growth. With irrigated crops, in particular when the drip system is used, fertilizer can be effectively delivered to the rooting zone during crop growth.

What Should You Do?

After reading the discussion above you may be somewhat bewildered by the different procedures and ways of expressing results, as well as the different recommendation approaches. The fact is that it is bewildering! Our general suggestions of how to deal with these complex issues are:

1. Send your soil samples to a lab that uses tests evaluated for the soils and crops of your state or region. Continue using the same lab or another that uses the same procedures and recommendation system.

2. If you're growing low value-per-acre crops (wheat, corn, soybeans, etc.), be sure that the recommendation system used is based on the sufficiency approach. This system usually results in lower fertilizer rates and higher economic returns for low value crops. [It is not easy to find out what system a lab uses. Be persistent and you will get to a person that can answer your question.]

3. Dividing the same sample in two and sending it to two labs may result in confusion. You will probably get different recom-

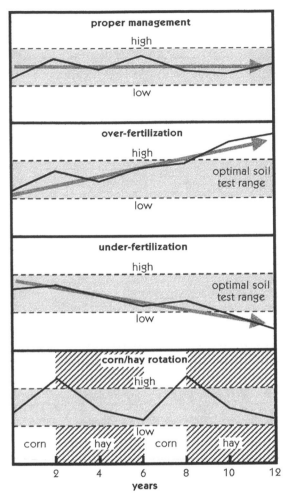

Figure 19.3 Soil test phosphorus and potassium trends under different fertility management regimes. Modified from The PennState Agronomy Guide, 1999.

two different labs, with one going to your state-testing laboratory. In general, the recommendations from these labs call for less, but enough, fertilizer. If growing crops over large acreage, set up a demonstration or experiment in one field where you apply the fertilizer recommended by each lab over long strips and see if there is any yield difference. A yield monitor for grain crops would be very useful for this purpose. If you've never set up a field experiment before, you should ask your extension agent for help, and might find the brochure *How to Conduct Research on Your Farm or Ranch* of use (see *Sources* at end of chapter).

4. Keep a record of soil tests for each field, so that you can track changes over the years (figure 19.3). If records show a build up of nutrients to high levels, reduce nutrient applications. If you're drawing nutrient levels down too low, start applying fertilizers or off-farm organic nutrient sources. In some rotations, such as the corn-corn-4 years of hay shown at the bottom of figure 19.3, it makes sense to build up nutrient levels during the corn phase and draw them down during the hay phase.

SOIL TESTING FOR N

Soil samples for nitrogen tests are usually taken at a different time and using a different method than for the other nutrients (which are typically sampled to plow depth in the fall or spring). Before the mid-1980s, there was no reliable soil test for N availability in the humid regions of the US.

The nitrate test commonly used for corn in humid regions was developed during the 1980s in Vermont. It is usually called the Pre-Sidedress

mendations and it won't be easy to figure out which is better for you, unless you are willing to do a comparison of recommendations. In most cases you are better off staying with the same lab and learning how to fine-tune recommendations for your farm. However, if you are willing to experiment, you can send duplicate samples to

Nitrate Test (PSNT), but also goes under other names (see box). All of these names refer to the same test — a soil sample is taken to 1 foot depth, when corn is between 6 inches and 1 foot tall. The original idea behind the test was to wait as long as possible before sampling, because soil and weather conditions in the early growing season may reduce or increase N availability for the crop later in the season. After the corn is 1 foot tall, it is difficult to get samples to a lab and back in time to apply any needed sidedress N fertilizer. The PSNT is now used on both field corn and sweet corn and research is underway in the northeastern U.S. to extend its use to pumpkins and cabbage. Although the PSNT is widely used, there are some situations, such as the sandy coastal plains soils of the deep south, where it is not very accurate.

Different approaches to using the PSNT work for different farms. In general, using the soil test allows a farmer to avoid adding excess amounts of "insurance fertilizer." Two contrasting examples follow:

- **For farms using rotations with legume forages and applying animal manures regularly (so there's a lot of active soil organic matter),** the best way to use the test to your advantage is to apply only the amount of manure necessary to provide sufficient N to the plant. The PSNT will indicate whether or not you need to side-dress any additional N fertilizer. It will also show you whether you've done a good job of estimating N availability from manures.

- **For farms growing cash grains without using legume cover crops,** it's best to apply a conservative amount of fertilizer N before planting and then use the test to see if more is needed. This is especially important in regions where rainfall cannot always be relied upon to quickly bring fertilizer into contact with roots. The PSNT test provides a backup and allows the farmer to be more conservative with preplant applications, knowing that there is a way to make up any possible deficit.

Other nitrogen soil tests. In the drier parts of the country, a nitrate soil test, requiring samples to 2 feet or more, has been used with success since the 1960s. The deep-soil samples can be taken in the fall or early spring, before the growing season, because of low leaching and denitrification losses and low levels of active organic matter (so hardly any nitrate is mineralized from organic matter). Soil samples can also be taken at the same time for analysis for other nutrients and pH. A few states in the upper Midwest offer a preplant nitrate test, which calls for sampling to 2 feet in the spring.

SOIL TESTING FOR P

Soil test procedures for phosphorus are different than for nitrogen. When testing for phosphorus, the soil is usually sampled to plow depth at a different time — in the fall or in the early spring before tillage — and the sample is usually analyzed for phosphorus, potassium, sometimes other nutrients (such as calcium, magnesium, and micronutrients) and pH. The methods used to estimate available P vary from region to region, and sometimes, from state to state within a region (Table 19.1). Although the relative test value for a given soil is usually similar when using different soil tests (for example, a high P-testing soil by one procedure is generally also high by another procedure), the actual numbers can be different (table 19.2).

The various soil tests for P take into account a large portion of the available P contained in recently applied manures and the amount that will become available from the soil minerals. However, if there is a large amount of active organic matter in your soils from crop residues or manure additions made in previous years, there may well be more available P for plants than indicated by soil test. (On the other hand, the PSNT reflects the amount of N that may become available from decomposing organic matter.)

TABLE 19.1
Phosphorus Soil Tests Used in Different Regions.

REGION	SOIL TEST SOLUTIONS USED FOR P
Arid and semi-arid Midwest, West and Northwest	Olsen AB-DTPA
Humid Midwest, mid-Atlantic, Southeast, and eastern Canada	Mehlich 3
North Central and midwest	Bray 1 (also called Bray P-1 or Bray-Kurtz P)
Southeast and mid-Atlantic	Mehlich 1
Northeast (New York and most of New England), some labs in Idaho and Washington	Morgan or modified-Morgan

—MODIFIED FROM ALLEN ET AL. (1994)

TABLE 19.2
Interpretation Ranges for Different P Soil Tests

	LOW	OPTIMUM	HIGH	VERY HIGH
Olsen	0 to 7	7 to 15	15 to 21	>21
Morgan	0 to 4	4 to 7	7 to 20	>20
Bray 1 (Bray P-1)	0 to 15	15 to 24	24 to 31	>31
Mehlich 1	0 to 25	25 to 50		>50
Mehlich 3	0 to 15	15 to 24	24 to 31	>31
AB—DTPA (for irrigated crops)	0 to 7	8 to 12	12 to15	>15

Note: units are in parts per million phosphorus (ppm P) and ranges used for recommendations may vary from state to state.

TESTING SOILS FOR ORGANIC MATTER

A word of caution when comparing your soil test organic matter levels with those discussed in this book. If your laboratory reports organic matter as "weight loss" at high temperature, the numbers may be higher than if the lab uses the traditional wet chemistry method. A soil with 3 percent organic matter by wet chemistry might have a weight-loss value of between 4 and 5 percent. Most labs use a correction factor to approximate the value you would get by using the wet chemistry procedure. Although either method can be used to follow changes in your soil, when you compare soil organic matter of samples run in different laboratories, it's best to make sure the same methods were used.

There is now a laboratory that will determine various forms of living organisms in your soil. Although it costs quite a bit more than traditional testing for nutrients or organic matter, you can find out the amount (weight) of fungi and bacteria in a soil, as well as analysis for other organisms. (See the Resources section at the back of the book for laboratories that run tests in addition to basic soil fertility analysis.)

INTERPRETING SOIL TEST RESULTS

Below are five soil test examples, including discussion about what they tell us and the types of practices that should be followed. Suggestions are provided for conventional farmers and organic producers. These are just suggestions — there are other satisfactory ways to meet the needs for crops growing on these soils. The soil tests were run by different procedures, to give

Unusual Soil Tests?

From time to time we've come across unusual soil test results. A few examples and their typical causes are given below.

Very high phosphorus levels. High poultry or other manure application over many years.

Very high salt concentration in humid region. Recent application of large amounts of poultry manure or immediately adjacent to road where de-icing salt was used.

Very high pH and high calcium levels, relative to potassium and magnesium. Large amounts of lime-stabilized sewage sludge were used.

Very high calcium levels given the soil's texture and organic matter content. Using an acid solution, such as the Morgan, Mehlich 1, or Mehlich 3, to extract soils containing free limestone causes some of the lime to dissolve, giving artificially high calcium test levels.

examples from around the US. Interpretations for a number of commonly used soil tests — relating test levels to general fertility categories — are given later in the chapter (tables 19.3 and 19.4). Many labs estimate the cation exchange capacity that would exist at pH 7 (or even higher). Because we feel that the soil's current CEC is of most interest (see chapter 18), the CEC is estimated by summing the exchangeable bases. The more acidic a soil, the greater the difference between its current CEC and the CEC it would have near pH 7.

Following the five soil tests is a section on modifying recommendations for particular situations.

— SOIL TEST #1 —
(New England)

Field name: North
Sample date: September (PSNT sample taken the following June)
Soil type: loamy sand
Manure added: none
Cropping history: mixed vegetables
Crop to be grown: mixed vegetables

What can we tell about soil #1 based on the soil test?

❑ It is too acidic for most agricultural crops, so lime is needed.
❑ Phosphorus is low, as are potassium, magnesium, and calcium. All should be applied.

❑ This low organic matter soil is probably also low in active organic matter (indicated by the low PSNT test, see table 19.4a) and will need an application of nitrogen. (The PSNT is done during the growth of the crop, so it is difficult to use manure to supply extra N needs indicated by the test.)
❑ The coarse texture of the soil is indicated by the combination of low organic matter and low CEC.

General recommendations:

1. Apply dolomitic limestone, if available, in the fall at about 2 tons/acre (and work it into the soil and establish a cover crop if possible). This will take care of the calcium and magnesium needs at the same time the soil's pH is increased.

Soil Test #1 Report Summary*

	LBS./ACRE	PPM	SOIL TEST CATEGORY	RECOMMENDATION SUMMARY
P	4	2	low	50–70 lbs. P_2O_5/acre
K	100	50	low	150–200 lbs. K_2O/acre
Mg	60	30	low	lime (see below)
Ca	400	200	low	lime (see below)
pH	5.4			2 tons dolomitic limestone/acre
CEC**	1.4 me/100g			
OM	1%			add organic matter: compost, cover crops, animal manures
PSNT		5	low	sidedress 80–100 lbs. N/acre

* Nutrients extracted by modified Morgan's Solution (see table 19.3a for interpretations).

** CEC by sum of bases. The estimated CEC would probably double if "exchange acidity" were determined and added to the sum of bases.

2. Because no manure is to be used after the test was taken, broadcast significant amounts of phosphate (probably around 50 to 70 lbs. phosphate (P_2O_5)/acre) and potash (around 150 to 200 lbs. potash (K_2O)/acre). Some phosphate and potash can also be applied in starter fertilizer (band applied at planting). Usually N is also included in starter fertilizer, so it might be reasonable to use about 300 lbs. of a 10-10-10 fertilizer, which will apply 30 lbs. of N, 30 lbs. of phosphate, and 30 lbs. of potash per acre. If that rate of starter is to be used, then broadcast 400 lbs. per acre of a 0-10-30 bulk blended fertilizer. The broadcast plus the starter will supply 30 lbs. of N, 70 lbs. of phosphate, and 150 lbs. of potash per acre.

3. If only calcitic (low magnesium) limestone is available, use Sul-Po-Mag as the potassium source in the bulk blend to help supply magnesium.

4. Nitrogen should be sidedressed at around 80 to 100 (or more) lbs./acre for N-demanding crops, such as corn or tomatoes. About 300 lbs. of ammonium nitrate or 220 lbs. of urea per acre will supply 100 lbs. of N.

5. Use various medium-to-long-term strategies to build up soil organic matter, including use of cover crops and animal manures.

Most of the nutrient needs of crops on this soil could have been met by using about 20 tons wet weight of solid cow manure/acre or its equivalent. It is best to apply it in the spring, before planting. If the manure had been applied, the PSNT test would probably have been quite a bit higher, perhaps around 25 ppm.

Recommendations for organic producers:

1. Use dolomitic limestone to increase the pH (as recommended for the conventional farmer above).

2. Apply 2 tons/acre of rock phosphate, or about 5 tons of poultry manure for phosphorus, or — better yet — a combination of 1 ton rock phosphate and 2½ tons of poultry manure. If the high level of rock phosphate is applied, it should supply some phosphorus for a long time, perhaps a decade.

3. If the poultry manure is used to raise the phosphorus level, add 2 tons of compost per acre to add some longer lasting nutrients and humus. If rock phosphate is used to supply phosphorus, then use livestock manure and compost (to add N, potassium, magnesium, and some humus).

4. Establish a good rotation with soil-building crops and cover crops.

5. Care is needed with manure use. Although the application of uncomposted manure is allowed by organic certifying organizations, there are restrictions. For example, three to four months may be needed between application of uncomposted manure and either harvest of root crops or planting of crops that accumulate nitrate, such as leafy greens or beets. A two-month period may be needed between uncomposted manure application and harvest of other food crops.

— SOIL TEST #2 —

(Pennsylvania, New York)

Field name: Smith upper

Sample date: November (no sample for PSNT will be taken)

Soil type: silt loam

Manure added: none this year (some last year)

Cropping history: legume cover crops used routinely

Crop to be grown: corn

What can we tell about soil #2 based on the soil test?

❏ The high pH indicates that this soil does not need any lime.

❏ Phosphorus is high, as are potassium, magnesium, and calcium (see table 19.3d).

❏ The organic matter is very good for a silt loam.

❏ There was no test done for nitrogen, but this soil probably supplies a reasonable amount of N for crops, because the farmer uses legume cover crops and allows them to produce a large amount of dry matter.

General recommendations:

1. Continue building soil organic matter.

2. No phosphate, potash, or magnesium needs to be applied. The lab that ran this soil test recommended using 38 lbs. potash (K_2O) and 150 lbs. of magnesium (MgO) per acre. However, with a high K level, 180 ppm (about 8 percent of the CEC) and a high Mg, 137 ppm (about 11 percent of the CEC), there is a very low likelihood of any increase in yield or crop quality from adding either element.

3. Nitrogen fertilizer is probably needed in only small to moderate amounts (if at all), but we

Soil Test #2 Report Summary*

	LBS./ACRE	PPM	SOIL TEST CATEGORY	RECOMMENDATION SUMMARY
P	174	87	high	none
K	360	180	high	none
Mg	274	137	high	none
Ca	3880	1940	high	none
pH	7.2			no lime needed
CEC	11.7 me/100g			
OM	3%			add organic matter: compost, cover crops, animal manures
N	No N soil test			little to no N needed

* Soil sent to a commercial lab. P using Bray-1 solution. This is probably the equivalent of over 20 ppm by using Morgan or Olsen procedures. Other nutrients extracted with pH 7 ammonium acetate (see table 19.3d).

need to know more about the details of the cropping system or run a nitrogen soil test to make a more accurate recommendation.

Recommendations for organic producers:

1. A good rotation with legumes will provide nitrogen for other crops.

— SOIL TEST #3 —
(Humid Midwest)

Field name: #12
Sample date: December (no sample for PSNT will be taken)
Soil type: clay (somewhat poorly drained)
Manure added: none
Cropping history: continuous corn
Crop to be grown: corn

What can we tell about soil #3 based on the soil test?

❑ The high pH indicates that this soil does not need any lime.
❑ Phosphorus and potassium are low.
❑ The organic matter is relatively high. However, considering that this is a somewhat poorly drained clay, it probably should be even higher.

Soil Test #3 Report Summary*

	LBS./ACRE	PPM	SOIL TEST CATEGORY	RECOMMENDATION SUMMARY
P	20	10	very low	30 lbs. P_2O_5/acre
K	58	29	very low	200 lbs. K_2O/acre
Mg	138	69	high	none
Ca	3168	4084	high	none
pH	6.8			no lime needed
CEC	21.1 me/100g			
OM	4.3%			rotate to forage legume crop
N	No N soil test			100-130 lbs. N/acre

*all nutrients determined using Mehlich 3 solution (see table 19.3c).

❏ About half of the CEC is probably due to the organic matter with the rest probably due to the clay.

❏ Low potassium indicates that this soil has probably not received high levels of manures recently.

❏ There was no test done for nitrogen, but given the field's history of continuous corn and little manure, there is probably a need for nitrogen. A low amount of active organic matter that could have supplied nitrogen for crops is indicated by past history (the lack of rotation to perennial legume forages and lack of manure use) and the moderate percent organic matter (considering that it is a clay soil).

General recommendations:

1. This field should probably be rotated to a perennial forage crop.

2. Phosphorus and potassium are needed. Probably around 30 lbs. of phosphate (P_2O_5) and 200 or more lbs. of potash (K_2O) applied broadcast, preplant, if a forage crop is to be grown. *If* corn will be grown again, all of the phosphate and 30 to 40 lbs. of the potash can be applied as starter fertilizer at planting. Although magnesium, at about 3 percent of the effective CEC, would be considered low by relying exclusively on a basic cation ratio saturation recommendation system, there is little likelihood of an increase in crop yield or quality by adding magnesium.

3. Nitrogen fertilizer is probably needed in large amounts (100 to 130 lbs./acre) for high N-demanding crops, such as corn. If no in-season soil test (like the PSNT) is done, some preplant N should be applied (around 50 lbs./acre), some in the starter band at planting (about 15 lbs./acre) and some side-dressed (about 50 lbs.).

4. One way to meet the needs of the crop is as follows:

    a) broadcast 500 lbs. per acre of an 11-0-44 bulk blended fertilizer;

    b) use 300 lbs. per acre of a 5-10-10 starter; and

    c) sidedress with 150 lbs. per acre of ammonium nitrate.

This will supply approximately 120 lbs. of N, 30 lbs. of phosphate and 210 lbs. of potash.

Recommendations for organic producers:

1. 2 tons/acre of rock phosphate (to meet P needs) *or* about 5 to 8 tons of poultry manure (which would meet both phosphorus and nitrogen needs), or a combination of the two (1 ton rock phosphate and 3 to 4 tons of poultry manure).

2. 400 lbs. of potassium sulfate per acre broadcast preplant. (If poultry manure is used to meet phosphorus and nitrogen needs, use only 200 to 300 lbs. of potassium sulfate per acre.)

3. Care is needed with manure use. Although the application of uncomposted manure is allowed by organic certifying organizations, there are restrictions. For example, three to four months may be needed between application of uncomposted manure and either harvest of root crops or planting of crops that accumulate nitrate, such as leafy greens or beets. A two-month period may be needed between uncomposted manure application and harvest of other food crops.

— SOIL TEST #4 —
(Alabama)

Field name: River A
Sample date: October
Soil type: sandy loam
Manure added: none
Cropping history: continuous cotton
Crop to be grown: cotton

What can we tell about soil #4 based on the soil test?

❑ With a pH of 6.5, this soil does not need any lime.
❑ Phosphorus is very high, and potassium and magnesium are sufficient.
❑ Magnesium is high, compared with calcium (Mg occupies over 26 percent of the CEC).
❑ The low CEC at pH 6.5 indicates that the organic matter content is probably around 1 to 1.5 percent.

General recommendations:

1. No phosphate, potash, magnesium, or lime is needed.
2. Nitrogen should be applied, probably in a split application totaling about 70 to 100 lbs. N/acre.
3. This field should be rotated to other crops and cover crops used regularly.

Recommendations for organic producer:

1. Although poultry or dairy manure can meet the crop's needs, that means applying phosphorus on an already high-P soil. If there is no possibility of growing an overwinter legume cover crop (see below), then about 15 to 20 tons of bedded dairy manure (wet weight) should be sufficient.
2. If time permits, this soil can use a high-N producing legume cover crop, such as hairy vetch or crimson clover, to provide nitrogen to cash crops.

*Soil Test #4 Report Summary

	LBS./ACRE	PPM	SOIL TEST CATEGORY	RECOMMENDATION SUMMARY
P	102	51	very high	none
K	166	83	high	none
Mg	264	132	high	none
Ca	1158	579		none
pH	6.5		moderate	no lime needed
CEC	4.2 me/100 g			
OM	not requested			use legume cover crops, consider crop rotation
N	No N soil test			70–100 lbs. N/acre

*all nutrients determined using Mehlich 1 solution (see table 19.3b).

3. Develop a good rotation so that all the needed nitrogen will be supplied to non-legumes between the rotation crops and cover crops.

4. Although the application of uncomposted manure is allowed by organic certifying organizations, there are restrictions when growing food crops. Check with the person doing your certification to find out what restrictions apply to cotton.

—SOIL TEST #5—
(Semi-arid Great Plains)

Field name: Hill
Sample date: April
Soil type: silt loam
Manure added: none indicated
Cropping history: not indicated
Crop to be grown: corn

What can we tell about soil #5 based on the soil test?

❑ The pH of 8.1 indicates that this soil is most likely calcareous.
❑ Phosphorus is low, there is sufficient magnesium, and potassium is very high.
❑ Although calcium was not determined, there will be plenty in a calcareous soil.

Soil Test #5 Report Summary*

	LBS./ACRE	PPM	SOIL TEST CATEGORY	RECOMMENDATION SUMMARY
P	14	7	low	20–40 lbs. P_2O_5
K	716	358	very high	none
Mg	340	170	high	none
Ca	not determined			none
pH	8.1			no lime needed
CEC	not determined			
OM	1.8%			use legume cover crops, consider rotation to other crops that produce large amounts of residues
N	5.8 ppm			170 lbs. N/acre

*K and Mg extracted by neutral ammonium acetate, P by Olsen solution (see table 19.3d).

❏ The organic matter at 1.8 percent is low for a silt loam soil.
❏ The nitrogen test indicates a low amount of residual nitrate (table 19.4b) and, given the low organic matter level, a low amount of N mineralization is expected.

General recommendations:

1. No potash, magnesium, or lime is needed.
2. About 170 lbs. of N/acre should be applied. Because of the low amount of leaching in this region, most can be applied pre-plant, with perhaps 30 lbs. as starter (applied at planting). Using 300 lbs. per acre of a 10-10-0 starter would supply all P needs (see below) as well as give some N near the developing seedling. Broadcasting and incorporating 300 lbs. of urea or 420 lbs. of ammonium nitrate will provide 140 lbs. of N.
3. About 20 to 40 lbs. of phosphate (P_2O_5) is needed per acre. Apply the lower rate as starter, because localized placement results in more efficient use by the plant. If phosphate is broadcast, apply at the 40 lb rate.
4. The organic matter level of this soil should be increased. This field should be rotated to other crops and cover crops used regularly.

Recommendations for organic producers:

1. Because rock phosphate is so insoluble in high pH soils, it would be a poor choice for adding P. Poultry (about 6 tons per acre) or dairy (about 25 tons wet weight per acre) manure can be used to meet the crop's needs for both N and P. However, that means applying more P than is needed, plus a lot of potash (which is already at very high levels).
2. A long-term strategy needs to be developed to build soil organic matter — better rotations, use of cover crops, and importing organic residues onto the farm.
3. Care is needed with manure use. Although the application of uncomposted manure is allowed by organic certifying organizations, there are restrictions. For example, three months may be needed between application of uncomposted manure and either harvest of root crops or planting of crops that accumulate nitrate, such as leafy greens or beets. A two-month period may be needed between uncomposted manure application and harvest of other food crops.

TABLE 19.3
Soil Test Categories for Various Extracting Solutions

A.
Modified Morgan's Solution (Vermont)

CATEGORY	LOW	MEDIUM	OPTIMUM	HIGH	EXCESSIVELY HIGH
PROBABILITY OF RESPONSE TO ADDED NUTRIENT	VERY HIGH	HIGH	LOW	VERY LOW	
Available P (ppm)	0–2	2–4	4–7	7–20	>20
K (ppm)	0–50	51–100	101–130	131–160	>160
Mg (ppm)	0–35	35–50	51–100	>100	–

B.
Mehlich 1 Solution (Alabama)*

CATEGORY	VERY LOW	LOW	MEDIUM	HIGH	EXCESSIVELY HIGH
PROBABILITY OF RESPONSE TO ADDED NUTRIENT	VERY HIGH	HIGH	LOW	VERY LOW	
Available P (ppm)	0–12	13–25	26–50	51–125	>125
K (ppm)	0–45	46–90	91–180	>180	
Mg (ppm)**	0–25	26–50	>50		
Ca for tomatoes (ppm)***	0–150	151–250	>250		

  * From Procedures Used by State Soil Testing Laboratories in the Southern Region of the United States, 1998.
 ** for corn, legumes, and vegetables on soils with CECs greater than 4.6 me/100g
*** for corn, legumes, and vegetables on soils with CECs from 4.6 to 9.0 me/100g

C.
Mehlich 3 Solution (North Carolina).*

CATEGORY	VERY LOW	LOW	MEDIUM	HIGH	EXCESSIVELY HIGH
PROBABILITY OF RESPONSE TO ADDED NUTRIENT	VERY HIGH	HIGH	LOW	VERY LOW	
Available P (ppm)	0–12	13–25	26–50	51–125	>125
K (ppm)	0–17	18–44	45–87	88–174	>175
Mg (ppm)**	0–30	31–60	>60		

 * From Procedures Used by State Soil Testing Laboratories in the Southern Region of the United States, 1998.
** percent of CEC is also considered

D.
Neutral Ammonium Acetate Solution
for K and Mg and Olsen or Bray-1 for P (Nebraska (P and K), Minnesota (Mg))

CATEGORY	VERY LOW	LOW	MEDIUM	HIGH	EXCESSIVELY HIGH
PROBABILITY OF RESPONSE TO ADDED NUTRIENT	VERY HIGH	HIGH	LOW	VERY LOW	VERY LOW
P (Olson,ppm)	0–3	4–10	11–16	17–20	>20
P (Bray-1,ppm)	0–5	6–15	16–24	25–30	>30
K (ppm)	0–40	41–74	75–124	125–150	>150
Mg (ppm)	0–50		51–100	>101	

TABLE 19.4
Soil Test Catagories for Nitrogen Tests

A.
Presidress Nitrogen Test (PSNT)*

CATEGORY	LOW	MEDIUM	OPTIMUM	HIGH	EXCESSIVE
PROBABILITY OF RESPONSE TO ADDED NUTRIENT	VERY HIGH	HIGH	LOW	VERY LOW	NONE
Nitrate-N (ppm)	0–10	11–22	23–28	29–35	>35

*Soil sample taken to 1 ft when corn is 6 to 12 inches tall.

B.
Deep (4ft) Nitrate Test (Nebraska)

CATEGORY	LOW	MEDIUM	OPTIMUM	HIGH	EXCESSIVE
PROBABILITY OF RESPONSE TO ADDED NUTRIENT	VERY HIGH	HIGH	LOW	VERY LOW	NONE
Nitrate-N (ppm)	0–6	7–15	15–18	19–25	>25

ADJUSTING A SOIL TEST RECOMMENDATION

Specific recommendations must be tailored to the crops you want to grow, as well as other characteristics of the particular soil, climate, and cropping system. Most soil test reports use information that you supply about manure use and previous crop to adapt a general recommendation for your situation. However, once you feel comfortable with interpreting soil tests, you may also want to adjust the recommendations for a particular need. What happens if you decide to apply manure after you sent in the form along with the soil sample? Also, you usually don't get credit for the nitrogen produced by

---

Making Adjustments to Fertilizer Application Rates

If information about cropping history, cover crops, or manure use is not provided to the soil testing laboratory, the report containing the fertilizer recommendation cannot take these factors into account. Below is an example of how you can modify the report's recommendations.

Past crop = corn
Cover crop = crimson clover, but small to medium amount of growth.
Manure = 10 tons of dairy manure that tested at 10 lbs. N, 3 lbs. of P_2O_5, and 9 lbs. of K_2O per ton. (A decision to apply manure was made after the soil sample was sent, so the recommendation could not take those nutrients into account.)

Worksheet for Adjusting Fertilizer Recommendations

	N	P_2O_5	K_2O
SOIL TEST RECOMMENDATION Accounts for contributions from the soil. Accounts for nutrients contributed from manure and previous crop only **if** information is included on form sent with soil sample.	120	40	140
CREDITS (Use only if not taken into account in recommendation received from lab.)			
Previous crop (already taken into account)	-0		
Manure (10 tons @ 6 lbs. N–2.4 lbs. P_2O_5–9 lbs. K_2O per ton, assuming that 60% of the nitrogen, 80% of the phosphorus and 100% of the potassium in the manure will be available this year.)	-60	-24	-90
Cover Crop (medium growth crimson clover)	-50		
TOTAL NUTRIENTS NEEDED FROM FERTILIZER	10	16	50

TABLE 19.5
Amounts of Available Nutrients from Manures and Legume Cover Crops

Legume Cover Crops[1]	N LBS./ACRE
Hairy vetch	70–140
Crimson clover	40–90
Red and white clovers	40–90
Medics	30–80

Manures[2]	N	P_2O_5	K_2O
	LBS. PER TON MANURE		
Dairy	6	4	10
Poultry	20	15	10
Hog	6	3	9

[1] Amount of available N varies with amount of growth.

[2] Amount of nutrients varies with diet, storage, and application method. Quantities given in this table are somewhat less than for the total amounts given in table 9.1.

legume cover crops because most forms don't even ask about their use. The amount of available nutrients from legume cover crops and from manures is indicated in table 19.5. Another common situation occurs because most farmers don't test their soil annually and the recommendations they receive are only for the current year. Under these circumstances, you need to figure out what to apply the next year or two, until the soil is tested again.

No single recommendation, based only on the soil test, makes sense for all situations. For example, your gut feeling might tell you that a test is too low (and fertilizer recommendations are too high). Let's say that you broadcast 100 lbs. N/acre before planting, but a high rate of N fertilizer is still recommended by the in-season nitrate test (PSNT), even though there wasn't enough rainfall to leach out nitrate or cause much loss by denitrification. In this case, you may not want to apply the full amount recommended.

Another example: a low potassium level in a soil test (let's say around 40 ppm) will certainly mean that you should apply potassium. But, how much should you use? When/how should you apply it? The answer to these two questions might be quite different on a low-organic matter, sandy soil where high amounts of rainfall normally occur during the growing season (in which case potassium may leach out if applied the previous fall, or early spring) versus a high-organic matter clay loam soil that has a higher CEC and will hold onto potassium added in the fall. This is the type of situation that dictates using labs whose recommendations are developed for soils and cropping systems in your home state or region. It also is an indication that you may need to modify a recommendation for your specific situation.

Sources
Allen, E.R., G.V. Johnson, and L.G. Unruh. 1994. Current approaches to soil testing methods: Problems and solutions. pp. 203–220. In *Soil Testing: Prospects for Improving Nutrient Recommendations* (J.L. Havlin et al., eds). Soil Science Society of America. Madison, WI.

Cornell Cooperative Extension. 2000. *Cornell Recommendations for Integrated Field Crop Production.* Cornell Cooperative Extension, Ithaca, NY.

Herget, G.W., and E. J. Penas. 1993. *New Nitrogen Recommendations for Corn.* NebFacts NF 93–111, University of Nebraska Extension. Lincoln, NE.

Jokela, B., F. Magdoff, R. Bartlett, S. Bosworth, and D. Ross. 1998. *Nutrient Recommendations for Field Crops in Vermont*. University of Vermont Extension. Brochure 1390. Burlington, VT.

Penas, E.J., and R.A. Wiese. 1987. *Fertilizer Suggestions for Soybeans*. NebGuide G87-859-A. University of Nebraska Cooperative Extension. Lincoln, NE.

Hanlon, E. (ed.). 1998. *Procedures Used by State Soil Testing Laboratories in the Southern Region of the United States*. Southern Cooperative Series Bulletin No. 190–Revision B. University of Florida. Immokalee, FL.

How to Conduct Research on Your Farm or Ranch. 1999. Available from SARE regional offices. Also available at: www.sare.org/san/htdocs/pubs

Recommended Chemical Soil Test Procedures for the North Central Region. 1998. North Central Regional Research Publication No. 221 (revised). Missouri Agricultural Experiment Station SB1001. Columbia, MO.

Rehm, G., 1994. *Soil Cation Ratios for Crop Production*. North Central Regional Extension Publication 533. University of Minnesota Extension. St. Paul, MN.

Rehm, G., M. Schmitt, and R. Munter. 1994. *Fertilizer Recommendations for Agronomic Crops in Minnesota*. University of Minnesota Extension. BU-6240-E. St Paul, MN.

The PennState Agronomy Guide. 1999. The Pennsylvania State University. University Station, PA.

PART THREE

Putting It All Together

20

How Good are Your Soils?
Assessing Soil Health

...the Garden of Eden, almost literally,
lies under our feet almost anywhere on the earth
we care to step. We have not begun to tap the
actual potentialities of the soil for producing crops.

—E.H. FAULKNER, 1943

By now, you should have some ideas about practices for increasing soil health on your farm. Most farmers know that soil health is important. However, when you work to improve your soils, how can you tell that they're actually getting any better? We're all used to taking soil samples and having them analyzed for available nutrients, pH, lime requirement, and total organic matter. Those tests are important pieces of information when used to adjust nutrient management practices. But the real issue is not how well a soil does in a lab test, but how well it does in the field. Does it do all the things we expect from a healthy soil? Of course, we want a soil to supply nutrients in adequate amounts but not in such excess that might cause plant health or environmental problems.

Does your soil also...
- Allow water to infiltrate easily during a downpour and drain afterward?
- Provide sufficient water to plants during dry spells?
    - Allow crops to fully develop healthy root systems?
    - Suppress root diseases and parasitic nematodes?

We can evaluate the quality of our soils in many different ways, just as we do with human health. We can assess human health with a variety of diagnoses or procedures, ranging from "you look a little pale today" to taking a person's temperature to doing a simple blood pressure test to computerized body imaging. For soils, we are limited in our ability to diagnose problems because we do not have the equivalent of the extensive medical knowledge base that is available for humans. We also have some additional challenges with soils. For example, a single blood sample will assess the entire human body, because the blood circulates rapidly through the entire vascular system. Soils do not function as a single organism, but as part of an ecosystem. Therefore, to obtain a good assessment of the soil's health, we need to make multiple observations — at different locations in a field and over a period of time.

A simple but very good place to start assessing a soil's health is to look at its general performance as you go about your normal practices. It's something like wondering about your own performance during the course of a day: Do you feel more tired than usual? Are you concentrating less well on tasks than you commonly do? Do you look paler than normal? These are indications that something isn't quite right. Likewise, there are signs of poor soil health you might notice as part of the normal process of growing crops:

- Are yields declining?
- Do crops perform as well as those on neighboring farms with similar soils?
- Do your crops quickly show signs of stress or stunted growth during wet or dry periods?
- Does the soil plow up cloddy, and is it difficult to prepare a good seedbed?
- Does the soil crust over easily?
- If you are no-tilling, is it difficult to get the planter to penetrate?

The next step should be a little more quantitative. In a few states, farmers and researchers have developed "soil health score cards" that you fill out for each field. Card choices may vary from state to state. The differences in soils and climates suggest that there is no uniform soil health card that can be used everywhere. Nor is there a magic number or index value for soil health. The goal of any evaluation is to help you make changes and improve your soil's health over time by identifying key limitations or problems. Perfect soil can't be created everywhere, but you want to help the land reach its fullest potential.

Whenever you try to become more quantitative in assessing your soils, you should be aware that measurements may naturally vary considerably within a field, or may change over the course of the year. For example, if you decide to evaluate soil hardness with a penetrometer, or by using a thin metal rod, your results depend on the soil moisture conditions at the time of measurement. If you use a penetrometer in June of a year with low early-season rainfall, you may find the soil quite hard. If you go back the next year following a wet spring, the soil may be much softer. You shouldn't conclude that your soil's health has dramatically improved because you really measured the effect of variable soil moisture on soil strength. Similarly, earthworms will be abundant in the plow layer when it's moist, but tend to go deeper into the soil during dry periods. This type of variability with time of year or climatic conditions should not discourage you from starting to evaluate your soil's health — just keep in mind the limitations of certain measurements. Also, you can take advantage of the fact that soil health problems tend to be more obvious during extreme conditions. It's a good idea to spend some extra time walking your fields and digging in the soil after extended wet or dry periods.

In the next section, we use ideas and expressions developed for soil health or soil quality cards in Maryland, Oregon, and Wisconsin.

SOIL HEALTH INDICATORS

The indicators are not discussed below in any special order — all are important to help you assess soil health as it relates to growing crops.

Soil testing is a very common way to assess your soil's health from a chemical perspective (see detailed discussion in chapter 19). This provides information on potential nutrient and pH imbalances. To get the most benefit from soil tests, sample soils frequently and keep good records. If you change soil management practices, look at soil test trends and evaluate whether your soil is improving. Evaluate whether your soil test values are remaining in the optimal range, without adding large amounts of fertilizers. Also, make sure that you do not end up with excessive nutrient levels, especially phosphorus and potassium, due to over-application of organic materials. If your soil test report includes information on cation exchange capacity (CEC), you should expect it to increase with increasing organic matter levels.

Soil color is an indicator of soil organic matter content, especially within the same general textural class. The darkness is an indicator of the amount of humus (see chapter 2) in the soil. We generally associate black soils with high quality. The Illinois color chart, relating color to organic matter content, is proving useful in other parts of the country. However, don't expect dramatic color change when you add organic matter; it may take years to notice a difference.

Soil organisms such as ants, termites, and earthworms are "ecosystem engineers" that aid the initial organic matter breakdown that allows other species to thrive. They are easily recognized, and their general abundance is strongly affected by temperature and moisture levels in the soil. Their presence is best assessed in mid-spring, after considerable soil warming, and in mid-fall, before the soils become cold. Spring and fall assessment are practical during moist, but not excessively wet, conditions. Just take a spadeful of soil from the surface layer and sift through it looking for bugs and worms. If the soil is teeming with life, this suggests that the soil is healthy. If few invertebrates are observed, then the soil may be a poor environment for soil life and organic matter processing is probably low. Earthworms are often used as an indicator species of soil biological activity (see table 20.1). The most common worm types, such as the garden and red worms, live in the surface layer when soils are warm and moist and feed on organic materials in the soil. The long nightcrawlers dig almost vertical holes that extend well into the subsoil, but they feed on residue at the surface. Look for the worms themselves as well as their casts (on the surface, for nightcrawlers) and holes to assess their presence. If you dig out a square foot of soil down to 1 foot depth and find 10 worms, the soil has a lot of earthworm activity.

Soil tilth and hardness can be assessed with an inexpensive penetrometer (the best tool), a tile finder, a spade, or a stiff wire (like those that come with wire flags). Tilth characteristics vary greatly during the growing season due to tillage, packing, settling (dependent on rainfall), crop canopy closure, and travel over the field to cultivate, apply pesticides, and harvest. It is, therefore, best to assess soil hardness several times during the growing season. If you do it only once, the best time is when the soil is moist, but not too wet — it should be in the friable state. Soil is generally considered too hard for root growth if the penetrometer resistance is greater that 300 psi. Note also whether the soil is harder below the plow layer. We cannot be very quantitative with tile finders and wire, but the soil is generally too hard when you cannot easily push them in. If you use a spade when soil is not too wet, evaluate how hard the soil is and also pay attention to the structure of the soil. Is the plow layer fluffy and does it mostly consist of granules of about quarter-inch size? Or does the soil dig up in large clumps? A good way to evaluate that is by lifting a spade full of soil and slowly turning it over. Does the soil break apart into granules or does it drop in large clumps? When you dig below the plow layer, take a spade full of soil and pull the soil clumps apart. They should generally come apart easily in well-defined aggregates of several inches in size. If the soil is compacted, it does not easily come apart in distinct units.

Root development also can be evaluated by digging and is best done when the crop is in its rapid phase of growth — generally during late spring. Have the roots properly branched and extend in all directions to their fullest potential for the particular crop? Look for obvious signs of problems: short stubby roots, abrupt changes in direction when hitting hard layers, signs of rot or other diseases. Make sure to dig deep enough to get a full picture of the rooting environment.

Crusting, ponding, runoff, and erosion can be observed from the soil surface. However, the extent of their occurrence depends on whether there was an intense rainstorm. The presence of these symptoms are a sign of poor soil health, but the lack of visible signs doesn't necessarily mean that the soil is in good health — it must rain hard for signs to occur. Try to get out into the field after heavy rainstorms, especially in the early growing season. Crusting is recognized by a dense layer at the surface, which may become hard after it dries. Ponding is recognized either directly when the water is still in a field depression, or afterwards by small areas where the soil has slaked (aggregates have disintegrated). Areas that were ponded often show cracks after drying. Slaked areas going down the slope are an indication that runoff and early erosion have occurred. When rills and gullies are present, a severe erosion problem is at hand. Another idea: put on your raingear and go out during a rainstorm (not during lightning, of course) and actually see runoff and erosion in action. Compare fields with different crops, management, or soil types. This might give you ideas about changes you can make to reduce runoff and erosion.

You also can easily get an idea about stability of soil aggregates, especially those near the surface. If the soil crusts readily, you already know the answer — the aggregates are not very stable and break down completely when wet. If the soil doesn't usually form a crust, you might take a sample of aggregates from the top 3 or 4 inches of soil from a number of different fields that seem to have different soil quality. Gently drop a number of aggregates from each field into separate cups that are half-filled with water — the aggregates should be covered with water. See if they hold up or if they break apart. You can swirl the water in the cups to see if that helps to break up the aggregates. Very turbid water indicates that the aggregates have broken down. If the water stays fairly clear, the aggregates are very stable.

TABLE 20.1
Qualitative Soil Health Indicators

Indicator	Best Assessed	Poor	Medium	Good
Earthworms	Spring/Fall. Good soil moisture.	0–1 worms in shovelful of top foot of soil. No casts or holes.	2–10 in shovelful. Few casts, holes, or worms.	10+ in top foot of soil. Lots of casts and holes in tilled clods. Birds behind tillage.
Organic Matter Color	Moist soil.	Topsoil color similar to subsoil color.	Surface color closer to subsoil color.	Topsoil clearly defined, darker than subsoil.
Organic Matter Residues	Anytime.	No visible residues.	Some residues.	Residues on most of soil surface.
Root Health	Late spring (rapid growth stage).	Few, thick roots. No subsoil penetration.	Off color (staining) inside root.	Roots fully branched and extended, reaching into subsoil. Root exterior and interior is white.
Subsurface Compaction	Pre-tillage or post harvest. Good soil moisture.	Wire breaks or bends when inserting flag.	Have to push hard, need fist to push flag in.	Flag goes in easily with fingers to twice the depth of plow layer.
Soil Tilth Mellowness Friability	Good soil moisture.	Looks dead. Like brick or concrete, cloddy. Either blows apart or hard to pull drill through.	Somewhat cloddy, balls up, rough pulling seedbed.	Soil crumbles well, can slice through, like cutting butter. Spongy when you walk on it.
Erosion	After heavy rainfall.	Large gullies over 2 inches deep joined to others, thin or no topsoil, rapid runoff the color of soil.	Few rills or gullies, gullies up to 2 inches deep. Some swift runoff, colored water.	No gullies or rills, clear or no runoff.
Water Holding Capacity	After rainfall. During growing season.	Plant stress two days after a good rain.	Water runs out after a week or so.	Holds water for a long period of time without signs of drought stress.
Drainage Infiltration	After rainfall.	Water lays for a long time, evaporates more than drains, always very wet ground.	Water lays for short period, eventually drains.	No ponding, no runoff, water moves through soil steadily. Soil not too wet, not too dry.
Crop Condition (How well it grows)	Growing season. Good soil moisture.	Problem growing throughout season, poor growth, yellow or purple color.	Fair growth, spots in field different, medium green color.	Normal healthy dark green color, excellent growth all season, across field.
pH	Anytime, but at same time of year each time.	Hard to correct for desired crop.	Easily correctable.	Proper pH for crop.
Nutrient Holding Capacity	Over a five-year period always at same time of year.	Soil tests dropping into "low" category.	Little change or slow down trend.	Soil tests trending up in relation to fertilizer applied and crop harvested but not into "very high" category.

—Modified from USDA. 1997.

The effects of soil health problems on general crop performance are most obvious during extreme conditions. That's why it is worthwhile to occasionally walk your fields during a wet period (when a number of rains have fallen or just after a long, heavy rain) or during an extended drought. During prolonged wet periods, poor soils often remain saturated for extended periods. The lack of aeration stunts the growth of the crop, and leaf yellowing indicates loss of available N by denitrification. This may even happen with high-quality soils if the rainfall is very excessive, but it is certainly aggravated by poor soil conditions. Dense, no-tilled soil may also show greater effects. Purple leaves indicate a phosphorus deficiency and are also often an indirect sign of stress on the crop. This may be related to soil health, but also can be brought on by other causes, such as cold temperatures.

Watch for stunted crop growth during dry periods and also look for the onset of drought stress — leaf curling or sagging leaves (depending on the crop type). Crops on soils that are in good health generally have delayed occurrence of drought stress. Poor soils, especially, may show problems when heavy rainfall, causing soil settling after tillage, is followed by a long drying period. Soils may hardset and completely stop crop growth under these circumstances. Extreme conditions are good times to look at crop performance and, at the same time, evaluate soil hardness and root growth.

Using the simple tools and observations suggested above, you can evaluate your soil's health. Soil health cards or soil quality books provide a place to record field notes and assessment information to allow you to compare changes that occur over the years. You also can make up your own assessment sheets.

OTHER TOOLS

More "scientific" measurements can be made: infiltration capacity, bulk density, the volume of large pores, soil strength, etc. However, making these measurements in a meaningful way is challenging and you should get a soil scientist or extension agent involved if you want to pursue more sophisticated measurements.

Soils also can be tested for their biological characteristics — for potentially harmful organisms (usually for specific species of nematodes) or, more broadly, for large organisms and microbiology. Laboratories that do these types of tests are listed in Resources on p.221.

SOURCE
U.S. Department of Agriculture (USDA). 1997. Maryland Soil Quality Assessment Book.

21

Putting It All Together

*... generally, the type of soil management that gives
the greatest immediate return leads to a deterioration
of soil productivity, whereas the type that provides
the highest income over the period of a generation leads
to the maintenance or improvement of productivity.*

—CHARLES KELLOGG, 1936

In this chapter, we'll give some guidelines on how you can promote high quality soils by adopting practices that maintain or increase soil organic matter, develop and maintain optimal soil physical conditions, and promote top-notch nutrient management. In earlier chapters of Part Two, we discussed many different ways to manage soils, crops, and residues, but we looked at each one as a separate strategy. In the real world, we need to combine a number of these approaches and use them together. In fact, each practice is related to, or impacts, other soil heath promoting practices. The real key is to modify and combine them in ways that make sense for your farm.

We hope that you don't feel as confused as the person on the left in the drawing on the next page. If the thought of making changes on your farm is overwhelming, you can start with only one or two practices that improve soil health. Not all of these suggestions are meant to be used on every farm.

Decisions on the farm need to support the economic bottom line. Research shows that the practices that improve soil health generally also improve the economics of the farm, in some cases dramatically. However, you need to consider the fact that the increased returns may not be immediate. After implementing new practices, soil health may improve at a slow rate and it may take a few years to see improved yields. A "learning period" is probably needed to make the new management practices work on your farm. Permit yourself to make a few mistakes. Changing management practices may involve an investment in new equipment. For example, changing tillage systems requires an investment in new till-

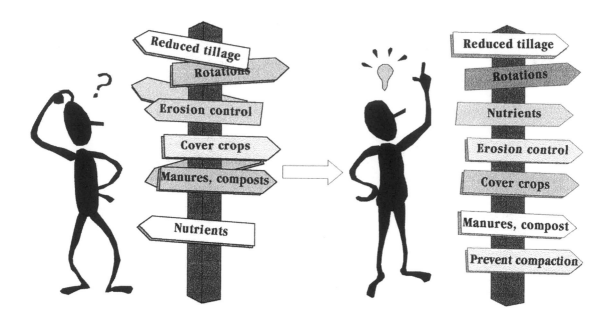

age tools and planters, and the bottom line may not improve immediately. For many farmers, the short-term limitations may keep them from making these changes, even though they are hurting the long-term viability of the farm. Big changes are probably best implemented at strategic times. For example, when you are ready to buy a new planter, consider a whole new approach to tillage as well. Also, take advantage of flush times, when you receive high prices for products, to invest in new management approaches. However, don't wait until that time to make decisions. Plan ahead, so you are ready to make the move at the right time.

GENERAL APPROACHES

There are many options for making soil management changes in different types of farming systems. Let's go over the general approaches that can be used for most types of agriculture. If

at all possible, use rotations that utilize grass, legume, or a combination of grass and legume sod crops, or crops with large amounts of residue as important parts of the system. Leave residues from annual crops in the field or, if you removed them for composting or to use as bedding for animals, return them to the soil as manure or compost. Use cover crops when soils would otherwise be bare to add organic matter, capture residual plant nutrients, and reduce erosion. Cover crops also help maintain soil organic matter in resource-scarce regions that lack possible substitutes to using crop residues for fuel or building materials.

Raising animals or having access to animal wastes from nearby farms gives you a wider choice of economically sound rotations. Rotations that include perennial forages make hay or pasture available for use by dairy and beef cows, sheep, and goats. In addition, on mixed crop-livestock farms, animal manures can be applied to cropland. It's easier to maintain organic matter on a diversified crop-and-livestock farm, where sod crops are fed to animals and manures returned to the soil. However, growing crops with high quantities of residues plus frequent use of green manures and composts from vegetative residues helps maintain soil organic matter even without animals.

You can maintain or increase soil organic matter more easily when you use reduced-tillage systems, especially no-till, instead of the conventional moldboard plow and disk system. The decreased soil disturbance under reduced tillage slows the rate of organic matter decomposition and helps to maintain a soil structure that allows rainfall to infiltrate rapidly. Leaving residue on the surface encourages the development of earthworm populations, which also improves soil structure. Compared with conventional tillage,

soil erosion is greatly reduced under minimum-tillage systems, which helps keep the organic matter and rich topsoil in place. Any practice that reduces soil erosion, such as contour tillage, strip-cropping along the contours, and terracing, also helps maintain soil organic matter.

Even if you use minimum-tillage systems to leave significant quantities of residue on the surface and decrease the severity of erosion, you also should use sound crop rotations. In fact, it may be more important to rotate crops when large amounts of residue remain on the surface. Decomposing residues harbor many insect and disease organisms. These problems may be worse in monoculture with no-till practices than with conventional tillage.

Test your soils regularly and apply lime and fertilizers only when they are needed. Testing soils every two or three years on each field is one of the best investments you can make. Make sure that you properly credit the N contribution of a decomposing sod or the N, P, and K contributions from manures. If you keep the report forms, or record the results, you will be able to follow the fertility changes over the years. Monitoring soil test changes will help you fine-tune your practices. Soil testing laboratories usually charge extra for an organic matter determination, but it's worth the money every few years just to track changes. Also, if you're interested in soil microorganisms, there is now a laboratory that can help you. In dry areas, salt accumulation may be a problem. You may need to use gypsum or other leaching salts. Also, maintain your pest scouting efforts and keep records over the years. This allows you to evaluate improvements in this area.

There is no substitute for taking a little time each year to observe your soils for such things

---

It's the Combination...

Farmers are learning that the combination of reduced tillage, cover crops, and better rotations can have a dramatic effect on their soil and the health of crops. They are finding that, by combining practices, they are reducing pest damage, improving soil tilth, vastly reducing runoff and erosion, increasing soil organic matter, and producing better crop growth. Each practice by itself is worthwhile. However, the greatest strengths and benefits are derived from combining a number of key practices.

For example, on the Groff farm (p. 145–146), it's the combination of good rotations, integrating livestock with crops, using no-till and cover crops that all work together to produce high-quality soil and crops.

as indications of compaction, the presence of earthworms, the health of roots or other indicators we discussed in chapter 20. The saying "The farmer's footprint is the best fertilizer" can be modified to "The farmer's footprint is the best path to improved soil health." If you don't already, begin to regularly observe and record the variability in crop yield across your fields. As equipment changes are made, you might consider buying a yield monitor that allows you to track yields on a field. Or, simply take the time to track production from the various sections of your fields that seem different. Compare your observations with your soil sampling plan, so you can be sure that the various areas within a field are receiving optimum management. Perhaps the hilltop or sideslope would benefit from additional manure or compost, while none is needed in other portions of the field.

WHAT MAKES SENSE ON *YOUR* FARM?

What makes sense on any individual farm depends on the soils, the climate, the nature of the farm enterprise itself and the surrounding region, potential markets, and the family's needs and goals. The specific details of implementing general management approaches depend primarily on the type of farm enterprise: grain or vegetable crops only, integrated crop-livestock, organic or not, etc.

Most grain crop farms export a lot of nutrients and are managed with a net loss of organic matter. However, these farms provide a great deal of flexibility in adopting alternative soil management systems because there is a wide range of equipment available for grain production systems. You can promote soil health easily with reduced-tillage systems, especially no-till and zone-till, instead of the conventional moldboard plow and disk system. Well-drained, coarse-textured soils are especially well adapted to no-till and zone-till systems, and the finer-textured soils do well with ridge-tillage or zone-tillage systems. Regardless of the tillage system that is used, you should try to travel on soils only when they're dry enough to resist compaction. However, managing no-till cropping on soils that are easily compacted is quite a challenge because there are few options to relieve compaction once it occurs. Maintaining controlled traffic zones or using some tillage to break up compacted layers may be necessary on such soils.

Even if you use minimum-tillage systems that leave significant quantities of residue on the surface and decrease the severity of erosion, you also should use sound crop rotations. Consider rotations that utilize grass, legume, or a combination of grass and legume perennial forage crops. Raising animals on what previously were exclusively crop farms, cooperating on rotations and manure management with a nearby farm, or growing forage crops for sale to a beef or dairy farm gives you a wider choice of economically sound rotations and at the same time helps to cycle nutrients better. Incorporating these innovations into a conventional grain farm often requires investment in new equipment and creatively looking for new markets for your products. There also are many opportunities to use cover crops on grain farms, even in reduced tillage systems.

Organic grain crop farms do not have the flexibility in soil management that conventional farms have. Tillage choices are limited because of the reliance on mechanical methods instead of herbicides to control weeds. On the positive side, organic farms already rely heavily on organic inputs through green and animal manures and composts to provide adequate nutrients to their crops. A well-managed organic farm usually uses many aspects of ecological soil management. However, erosion may remain a concern because many organic farms use clean and intensive tillage. It is important to think about reducing tillage intensity and perhaps invest in a better planter. New mechanical cultivators can generally handle higher residue and mulch levels and may still provide adequate weed control. Try to look into ways to increase surface cover, although this is a challenge without the use of chemical weed control. Alternatively, you should consider conventional erosion control practices, such as strip cropping, as they work well with rotations involving sod and cover crops.

Diversified crop-and-livestock farms have an inherent advantage for improving soil health.

Diversified crop-and-livestock farms have an inherent advantage for improving soil health. Crops can be fed to animals and manures returned to the soil, thereby providing a continuous supply of organic materials. For many livestock operations, perennial forage crops are a logical part of the cropping system, thereby reducing erosion potential and improving soil physical properties. Livestock-based farms also have some disadvantages. It is more difficult to adopt minimum tillage practices when sod crops are rotated with row crops, and the need to in-

corporate manure requires at least some type of tillage. You should still consider minimizing tillage by trying to inject the manure or chiseling it in, rather than plowing it under. Also, minimize soil pulverization by reducing secondary tillage and establishing the crops with no-tillage (or zone-tillage) planters.

Preventing soil compaction is important on many livestock-based farms. Manure spreaders are typically heavy and frequently go over the land at very unfavorable times, doing a lot of compaction damage. Think about ways to minimize this. In the spring, allow the fields to dry adequately (do the ball test) before taking spreaders out. If there is no manure storage, building a structure to hold it temporarily allows you to avoid the most damaging soil conditions.

Livestock farms require special attention to nutrient management, making sure that the organic nutrient sources are optimally used around the farm and that no negative environmental impacts occur. This requires a comprehensive look at all nutrient flows on the farm, finding ways to most efficiently use them, and preventing problems with excesses.

Soil quality management is especially difficult on vegetable farms. Many vegetable crops are sensitive to soil compaction and often pose greater challenges in pest management. These cropping systems, therefore, can greatly benefit from improved soil health. Most vegetable farms are not integrated with livestock production, and it is difficult to maintain a continuous supply of fresh organic matter. Bringing manure, compost, or other locally available sources of organic materials to the farm should be seriously considered. In some cases, vegetable farms can economically use manure from nearby livestock operations or swap land with them in a rotation. Farms near urban areas may benefit from leaves and grass clippings and municipal or food waste composts, which are increasingly becoming available. In such case, care should be taken to insure that the compost does not contain contaminants.

Vegetable cropping systems are generally well adapted to the use of cover crops because the main cropping season is generally shorter than those for grain and forage crops. There is usually sufficient time for growth of cover crops in the pre- or post-season to gain real benefits, even in colder climates. Using the cover crop as a mulch (or importing mulch materials from off the farm) appears to be a good system for certain fresh market vegetables, as it keeps the crop from direct contact with the ground, thereby reducing the potential for rot or disease.

The need to harvest crops during a very short period before quality declines — regardless of soil conditions — often results in severe compaction problems on large vegetable farms using large-scale equipment. Controlled traffic systems, including permanent beds, should be given serious consideration. Limiting compaction to narrow lanes and using other soil-building practices in between them is the best way to avoid severe compaction damage under those conditions.

THE FUTURE

Each of the farming systems discussed above has its limitations and opportunities for building better soils. Although there are ways to improve soil health in any system, the details may differ. Whatever crops you grow, when you creatively combine a reasonable number of practices that promote high quality soils, most of your farm's soil fertility problems should be solved along the way. The health and yield of your crops should improve. The soil will have more available nutrients, more water for plants to use, and better tilth. There should be fewer problems with diseases, nematodes, and insects. By concentrating on the practices that build high quality soils, you also will leave a legacy of land stewardship for your children and their children to inherit and follow.

Glossary

Alkaline soil. A soil with a pH above 7, containing more base than acid.

Allelopathic effect. The effect that some plants have in suppressing the germination or growth of other plants. The chemicals responsible for this effect are produced during the growth of a plant or during the decomposition of its residues.

Acid. A solution containing free hydrogen ions (H^+) or a chemical that will give off hydrogen ions into solution.

Acidic soil. A soil that as a pH below 7. The lower the pH, the more acidic is the soil.

Aggregates. The structures, or clumps, formed when soil minerals and organic matter are bound together with the help of organic molecules, plant roots, fungi, and clays.

Anion. A negatively charged element or molecule such as chloride (Cl^-) or nitrate (NO_3^-).

Ammonium (NH_4^+). A form of nitrogen that is available to plants and is produced in the early stage of organic matter decomposition.

Available nutrient. The form of a nutrient that a plant is able to use. Nutrients that the plant needs are commonly found in the soil in forms that the plant can't use (such as organic forms of nitrogen) and must be converted into forms that the plant is able to take into the roots and use (such as the nitrate form of nitrogen).

Ball test. A simple field test to determine soil readiness for tillage. A handful of soil is taken and squeezed into a ball. If the soil molds together, it is in the plastic state and too wet for tillage or field traffic. If it crumbles, it is in the friable state.

Base. Something that will neutralize an acid, such as hydroxide or limestone.

Beds. Small hilled-up, or raised, zones where crops (usually vegetables) are planted. They provide better-drained and warmer soil conditions. Similar to *ridges,* but are generally broader, and are usually shaped after conventional tillage has occurred.

Buffering. The slowdown or inhibition of changes. A substance that has the ability to buffer a solution is also called a buffer. Buffering can slow down pH changes by neutralizing acids or bases.

Bulk density. The mass of dry soil per unit volume. It is an indicator of the compactness of the soil.

Calcareous soil. A soil in which finely divided lime is naturally distributed, usually has a pH from 7 to slightly more than 8.

Cation. A positively charged ion such as calcium (Ca^{++}) or ammonium (NH_4^+).

Cation exchange capacity (CEC). The amount of negative charge that exists on humus and clays, allowing them to hold onto positively charged chemicals (cations).

Chelate. A molecule that uses more than one bond to attach strongly to certain elements such as iron (Fe^{++}) and zinc (Zn^{++}). These elements may later be released from the chelate and used by plants.

Coarse textured. Soil dominated by large mineral particles (sand size). May also include gravels. Previously called "light soil."

Colloid. A very small particle, with a high surface area, that can stay in a water suspension for a very long time. The colloids in soils, the clay and humus molecules, are usually found in larger aggregates and not as individual particles. These colloids are responsible for many of the chemical and physical properties of soils, including cation exchange capacity, chelation of micronutrients, and the development of aggregates.

Compost. Organic material that has been well decomposed by organisms under conditions of good aeration and high temperature, often used as a soil amendment.

Controlled traffic. Field equipment is restricted to limited travel or access lanes in order to reduce compaction on the rest of the field.

Conventional tillage. Preparation of soil for planting by using the moldboard plow followed by disking or harrowing. It usually breaks down aggregates, buries most crop residues and manures, and leaves the soil smooth.

Coulter. A fluted or rippled disk mounted on the front of a planter to cut surface crop residues and perform minimal soil loosening prior to seed placement. Multiple coulters are used on zone-till planters to provide a wider band of loosened soil.

Cover crop. A crop grown for the purpose of protecting the soil from erosion during the time of the year when the soil would otherwise be bare. It is sometimes called a green manure crop.

Crumb. A soft, porous, more-or-less round soil aggregate. Generally indicative of good soil tilth.

Crust. A thin, dense layer at the soil surface that becomes hard upon drying.

C:N ratio. The amount of carbon in a residue divided by the amount of nitrogen. A high ratio results in low rates of decomposition and can also result in a temporary decrease in nitrogen nutrition for plants, as micro-organisms use much of the available nitrogen.

Deep tillage. Tillage practices that loosen the soil at greater depths (usually greater than 12 inches) than during regular tillage.

Disk. An implement for harrowing, or breaking up, the soil. It is commonly used following a moldboard plow, but is also used by itself to break down aggregates, help mix fertilizers and manures with the soil, and smooth the soil surface.

Drainage. The process of soil water loss by percolation through the profile as a result of the gravitational force. Also: Removal of excess soil water through the use of channels, ditches, soil shaping and subsurface drain pipes.

Element. All matter is made up of elements, seventeen of which are essential for plant growth. Elements such as carbon, oxygen, and nitrogen combine to form larger molecules.

Erosion. Wearing away of the land surface by runoff water (*water* erosion), wind shear (*wind* erosion) or tillage (*tillage* erosion).

Field capacity. Water content of a soil following drainage by gravity.

Fine textured. Soils dominated by small mineral particles (silt and clay). Previously called "heavy soil."

Friable. Consistency status when soil crumbles instead of being molded when a force is applied.

Frost tillage. Tillage practices performed when a shallow (2–4 inches) frozen layer exists at the soil surface.

Full-field (full-width) tillage. Any tillage system that results in soil loosening over the entire width of the tillage pass. For example, moldboard plowing, chisel tillage, and disking.

Green manure. A crop grown for the main purpose of building up or maintaining soil organic matter. It is sometimes called a *cover crop.*

Heavy soil. Nowadays usually called "fine texture" soil, it contains a lot of clay and is usually more difficult to work than coarser texture soil. It normally drains slowly following rain.

Humus. The very well decomposed part of the soil organic matter. It has a high cation exchange capacity.

Infiltration. The process of water entering into the soil at the surface.

Inorganic. Chemicals that are **not** made from chains or rings of carbon atoms (for example, soil clay minerals, nitrate, and calcium).

Least-limiting water range. See optimum water range.

Legume. A group of plants including beans, peas, clovers, and alfalfa that forms a symbiotic relationship with nitrogen-fixing bacteria living in their roots. These bacteria help to supply plants with an available source of nitrogen.

Lignin. A substance found in woody tissue and in stems of plants that is difficult for soil organisms to decompose.

Lime, or limestone. A mineral that can neutralize acids and is commonly applied to acid soils, consisting of calcium carbonate ($CaCO_3$).

Loess soils. Soils formed from windblown deposits of silty and fine-sand-size minerals. They are easily eroded by wind and water.

Micronutrient. An element, such as zinc, iron, copper, boron, and manganese, needed by plants only in small amounts.

Microorganism. Very small and simple organism such as bacteria and fungi.

Mineralization. The process by which soil organisms change organic elements into the "mineral" or inorganic form as they decompose organic matter (for example, organic forms of nitrogen are converted to nitrate).

Moldboard plow. A commonly used plow that completely turns over the soil and incorporates any surface residues, manures, or fertilizers deeper into the soil.

Monoculture. Production of the same crop in the same field year after year.

Mycorrhizal relationship. The mutually beneficial relationship that develops between plant roots of most crops and fungi. The fungi help plants obtain water and phosphorus by acting like an extension of the root system and in return receive energy-containing chemical nutrients from the plant.

Nitrate (NO_3^-). The form of nitrogen that is most readily available to plants. It is the nitrogen form normally found in the greatest abundance in agricultural soils,

Nitrogen fixation. The conversion of atmospheric nitrogen by bacteria to a form that plants can use. A small number of bacteria, which include the rhizobia living in the roots of legumes, are able to make this conversion.

Nitrogen immobilization. The transformation of available forms of nitrogen, such as nitrate and ammonium, into organic forms that are not readily available to plants.

No-till. A system of planting crops without tilling the soil with a plow, disk, chisel, or other tillage implement.

Optimum tillage water range. The range of soil water contents in which plants do not experience stress from drought, high soil strength, or lack of aeration.

Organic. Chemicals that contain chains or rings of carbon connected to one another. Most of the chemicals in plants, animals, microorganisms, and soil organic matter are organic.

Oxidation. The combining of a chemical such as carbon with oxygen, usually resulting in the release of energy.

Penetrometer. A device measuring soil resistance to penetration, an indicator of the degree of compaction. Includes a cone-tipped metal shaft that is slowly pushed into the soil while the resistance force is measured.

Perennial forage crops. Crops such as grasses, legumes, or grass/legume mixtures that form complete soil cover (sods) and are grown for pasture or to make hay and haylage for animal feed.

pH. A way of expressing the acid status, or hydrogen ion (H^+) concentration, of a soil or a solution on a scale where 7 is neutral, less than 7 is acidic, and greater than 7 is basic.

Photosynthesis. The process by which green plants capture the energy of sunlight and use carbon dioxide from the atmosphere to make molecules needed for growth and development.

Plastic. State of a soil when it molds easily when a force is applied. Compare to friable.

Plastic limit. Water content of soil at the transition from the plastic to the friable state. Upper limit of soil moisture where tillage and field traffic do not result in excessive compaction damage.

Polyculture. Growth of more than one crop in a field at the same time.

PSNT. The Pre-Sidedress Nitrate Test is a soil test for nitrogen availability where the soil is sampled to 1 ft depth during the early crop growth.

Respiration. The biological process that allows living things to use the energy stored in organic chemicals. In this process, carbon dioxide is released as energy is made available to do all sorts of work.

Restricted tillage. Any tillage system that includes only limited and localized soil disturbance in bands where plant rows are to be established. For example, no-till, zone-till, strip-till and ridge-till. Compare with full-field tillage.

Rhizobia bacteria. Bacteria that live in the roots of legumes and have a mutually beneficial relationship with the plant. These bacteria fix nitrogen, providing it to the plant in an available form, and in return receive energy-rich molecules that the plant produces.

Ridge tillage. Crops are planted on top of a small ridge (usually 2–4 inches height) which is generally re-formed annually with a special cultivator.

Rotation effect. The crop-yield benefit from rotations that includes better nutrient availability, fewer pest problems, and better soil structure.

Runoff. Water lost by flow over the soil surface.

Saline soil. A soil that contains excess free salts, usually sodium and calcium chlorides.

Saturation, soil. When all soil pores are water-filled (a virtual absence of soil air).

Silage. A feed produced when chopped-up corn plants or wilted hay are put into air-tight storage facilities (silos) and partially fermented by bacteria. The acidity produced by the fermentation and the lack of oxygen help preserve the quality of the feed during storage.

Slurry (manure). A manure that is between solid and liquid. It flows slowly and has the consistency of something like a very thick soup.

Sod crops. Grasses or legumes such as timothy and white clover that tend to grow very close together and form a dense cover over the entire soil surface.

Sodic soil. A soil containing excess amounts of sodium. If it is not also saline, clay particles disperse and the soil structure may be poor.

Strip cropping. Growing two or more crops in alternating strips, usually along the contour or perpendicular to the prevailing wind direction.

Structure. The physical condition of the soil. It depends upon the amount of pores, the arrangement of soil solids into aggregates, and the degree of compaction.

Texture. A name that indicates the relative significance of a soil's sand, silt, and clay content. The term "coarse texture" means that a soil has a high sand content, while "fine texture" means that a soil has a high clay content.

Thermophilic bacteria. Bacteria that live and work best under high temperatures, around 110° to 140° F. They are responsible for the most intense stage of decomposition that occurs during composting.

Tillage. The mechanical manipulation of soil, generally for the purpose of creating soil loosening, a seedbed, controlling weeds, and incorporation of amendments. *Primary* tillage (moldboard plowing, chiseling) is a more rigorous practice, primarily for soil loosening and incorporation of amendments. *Secondary* tillage (disking, harrowing) is a less rigorous practice following primary tillage that creates a seedbed containing fine aggregates.

Tillage erosion. The downslope movement of soil through the action of tillage implements.

Tilth. The physical condition, or structure, of the soil as it influences plant growth. A soil with good tilth is very porous and allows rainfall to infiltrate easily, permits roots to grow without obstruction, and is easy to work.

Wilting point. The moisture content when a soil contains only water that is too tightly held to be available to plants.

Zone tillage. A restricted tillage system that establishes a narrow (4–6 inches width) band of loosened soil with surface residues removed. This is accomplished using multiple coulters and row cleaners as attachments on a planter. May also include a separate "zone-building" practice that provides deep, narrow ripping without significant surface disturbance. A modification of no tillage, generally better adapted to cold and wet soils.

Resources

General Information Sources

ATTRA (Appropriate Technology Transfer for Rural Areas), the sustainable farming information center funded by the U.S. Department of Agriculture, provides assistance, publications and resources, including sustainable soil management, cover crops and green manures, farm-scale composting, and nutrient cycling in pastures, free of charge. P.O. Box 3657, Fayetteville, AR 72702; (800) 346-9140; www.attra.org

The Alternative Farming Systems Information Center (AFSIC) of USDA's National Agricultural Library compiles bibliographies and resource lists on topics of current interest, such as *Soil Organic Matter: Impacts on Crop Production* QB 91-24, *Compost: Application and Use*, QB 97-01, *Legumes in Crop Rotations*, QB 94-38, and *Dairy Farm Manure Management*, QB 95-02. National Agricultural Library, 10301 Baltimore Ave., Beltsville, MD, 20705-2351; (301) 504-6559; www.nal.usda.gov/afsic

How to Conduct Research on Your Farm or Ranch, an informational bulletin from the Sustainable Agriculture Network (SAN), provides practical tips for laying out a research trial. On-farm research can help you evaluate new practices such as fertilizer rates or cover crop species. Available from SARE (see next page) or at: www.sare.org/san/htdocs/pubs/

Most state Cooperative Extension offices publish leaflets and booklets on manures, soil fertility, cover crops, and other subjects described in this book. Request a list of publications from your county extension office. A number of states also have sustainable agriculture centers that publish newsletters.

USDA's Sustainable Agriculture Research and Education (SARE) program studies and spreads information about sustainable agriculture via a nationwide grants program. SARE funds publications through its Sustainable Agriculture Network (SAN), and maintains a database of more than 1,600 projects. For information about publications, funded projects and how to apply for a grant, call (301) 405-3186 or visit www.sare.org.

The Sustainable Farming Connection website — managed by former staff members of *The New Farm* magazine — offers practical information to farmers through a diverse collection of resources and web links on soil health, cover crops, composts, and related topics. metalab.unc.edu/farming-connection/soilhlth/home.htm

The "Soil Biology Primer" presents an introduction to the living soil system for natural resource specialists, farmers, and others. This set of eight units describes the importance of soil organisms and the soil food web to soil productivity and water and air quality. 1-888-LAND-CARE; landcare@swcs. org.

Manures, Fertilizers, Tillage, and Rotations

Best Management Practices Series: Soil Management, Nutrient Management, and No-Till. Ontario Ministry of Agriculture and Rural Affairs with Agriculture and Agri-Food Canada. Provides practical information on these subjects to farmers and crop advisers. Available from Ontario Federation of Agriculture, Attn. Manager, BMP, 40 Eglinton Ave. E., 5th Floor, Toronto, Ontario, M4P 3B1, Canada.

"Crop Rotations in Sustainable Production Systems." Francis, C.A., and M.D. Clegg. 1990. pp. 107–122. In *Sustainable Agricultural Systems* (C.A. Edwards, R. Lal, P. Madden, R.H. Miller, and G. House, eds.). Soil and Water Conservation Society, 7515 NE Ankeny Road, Ankeny, Iowa, 50021; (515) 289-2331; www.swcs.org/f_publications.htm

The Farmer's Fertilizer Handbook. Cramer, Craig, and the editors of *The New Farm*. 1986. Regenerative Agriculture Association. Emmaus, PA. This handbook contains lots of very good information on soil fertility, soil testing, use of manures, and use of fertilizers.

Fertile Soil: A Growers Guide to Organic and Inorganic Fertilizers. Parnes, R. 1990. Fertile Ground Books, P.O. Box 2008, Davis, CA 95617; 800-540-0170.

Michigan Field Crop Ecology: Managing biological processes for productivity and environmental quality. 1998. Cavigelli, M.A., S.R. Deming, L.K. Probyn, and R. R. Harwood (eds.). Michigan State University Extension Bulletin E-2646. East Lansing, MI.

No-Till Vegetables: A Sustainable Way to Increase Profits, Save Soil and Reduce Pesticides. Steve Groff. This video covers the basics of sustainable no-till vegetable production, detailing methods to control weeds and improve soil using cover crops or plant residue on his 175-acre Cedar Meadow Farm. $21.95 plus $3 s/h to Cedar Meadow Farm, 679 Hilldale Rd., Holtwood, PA 17532; (717) 284-5152. www.cedarmeadowfarm .com

Soil Fertility and Organic Matter as Critical Components of Production Systems. Follett, R. F., J. W. B. Stewart, and C. V. Cole (eds.). 1987. SSSA Special Publication No. 19. Soil Science Society of America, American Society of Agronomy. Madison, WI.

Soils for Management of Organic Wastes and Wastewaters. Elliott, L.F., and F. J. Stevenson (eds.). 1977. Soil Science Society of America. Madison, WI.

Soil Management for Sustainability. Lal, R., and F. J. Pierce (eds.). 1991. Soil and Water Conservation Society, 7515 NE Ankeny Road, Ankeny, Iowa, 50021; (515) 289-2331; www.swcs.org/f_publications.htm

USDA Natural Resources Conservation Service/Soil Quality Institute - Agronomy Technical Notes Series. The NRCS technical note series provides an excellent introduction to cover crops, effect of conservation crop rotation on soil quality, effects of residue management & notill on soil quality, legumes and soil quality, and related topics. Free from NRCS Soil Quality Institute, 2150 Pammel Drive, Ames, IA 50011; (515) 294-4592. www.statlab.iastate.edu/survey/SQI/agronomy.shtml

Soils, Soil Organisms, and Composting

Ecology of Compost. Dindal, D. 1972. Office of News and Publications, 122 Bray Hall, SUNY College of Environmental Science and Forestry, 1 Forestry Drive, Syracuse, NY, 13210-2778; (315) 470-6644.

Effects of Conversion to Organic Agricultural Practices on Soil Biota. Werner, M.R., and D.L. Dindal. 1990. American Journal of Alternative Agriculture 5(1):24-32.

The Field Guide to On-Farm Composting. Dougherty, M. (ed.). 1999. NRAES-114. Natural Resource, Agriculture, and Engineering Service. Ithaca, NY. NRAES, 152 Riley Robb Hall, Cooperative Extension, Ithaca, NY, 14853-5701. www.nraes.org

NRCS Soil Quality Website. The Soil Quality Institute identifies soil quality research findings and practical technologies that help conserve and improve soil, and enhance farming, ranching, forestry, and gardening enterprises. USDA-NRCS Soil Quality Institute, 2150 Pammel Drive, Ames, Iowa, 50011; 515-294-4592; www.statlab.iastate.edu/survey/SQI/

The Nature and Properties of Soils. 12th ed. Brady, N.C., and R.R. Weil. 1999. Macmillan Publishing Co. New York, NY.

On Farm Composting. Rynk, R. (ed.). 1992. NRAES-54. Natural Resource, Agriculture, and Engineering Service. Ithaca, NY. Contact NRAES, 152 Riley Robb Hall, Cooperative Extension, Ithaca, NY 14853-5701 or www.nraes.org

The Pedosphere and Its Dynamics: A Systems Approach to Soil Science. University of Alberta, Canada. An award-winning website on soil science, www.pedosphere.com

Phytohormones in Soils: Microbial Production and Function. Frankenberger, Jr., W.T., and M. Arshad. 1995. Marcel Dekker, Inc. New York, NY.

The Rodale Book of Composting: Easy Methods for Every Gardener. Martin, D.L., and G. Gershuny (eds.). 1992. Rodale Press. Emmaus, PA.

Soil Microbiology: An Exploratory Approach. Coyne, M.S. 1999. Delmar Publishers. Albany, NY.

Soil Microbiology and Biochemistry. Paul, E.A., and F.E. Clark. 1989. Academic Press. San Diego, CA.

Cover Crops

University of California's SAREP (Sustainable Agriculture Research and Education Program) The UC-SAREP Cover Crops Resource Page provides access to a host of on-line and in-print educational materials, including the very informative UC-SAREP Cover Crop Database. www.sarep.ucdavis.edu/ccrop/

Cover Crops for Clean Water. Hargrove, W.L. (ed.). 1991. Soil and Water Conservation Society. 7515 NE Ankeny Road, Ankeny, Iowa, 50021; 515-289-2331; www.swcs.org/f_publications.htm

Green Manuring Principles and Practices. Pieters, A.J. 1927. John Wiley & Sons. New York, NY.

An oldie but goody. This is an out-of-print book that some readers may enjoy sifting through. It can be located in college libraries, or borrowed through Inter-Library Loan.

Managing Cover Crops Profitably, 2nd Edition. 1998. Sustainable Agriculture Network, Handbook Series, No. 3. USDA Sustainable Agriculture and Education Program. An excellent source of practical information about cover crops. $19 plus $3.95 s/h to Sustainable Agriculture Publications, Rm. 10, Hills Bldg., University of Vermont, Burlington, VT 05405-0082; www.sare.org

Northeast Cover Crop Handbook. Sarrantonio, M. 1997. Soil Health Series, Rodale Institute. Kutztown, PA.

The Role of Legumes in Conservation Tillage Systems. Power, J.F. (ed.). 1987. Soil & Water Conservation Society, 7515 NE Ankeny Road, Ankeny, Iowa, 50021; 515-289-2331; www.swcs.org/f_publications.htm

Dynamics and Chemistry of Organic Matter

Building Soils for Better Crops. 1st Edition. Magdoff, F. 1992. University of Nebraska Press, Lincoln, NE. The last two chapters of the first edition contain information on the chemistry and dynamics of soil organic matter.

Humic, Fulvic, and Microbial Balance: Organic Soil Conditioning. Jackson, William R. 1993. Jackson Research Center. Evergreen, CO.

Humus Chemistry: Genesis, Composition, Reactions. 2nd Edition. Stevenson, F.J. 1994. Wiley & Sons. New York, NY.

"Soil carbon dynamics and cropping practices." Lucas, R.E., J.B. Holtman, and J.L. Connor. 1977. pp. 333–351. In *Agriculture and Energy* (W. Lockeretz, ed.). Academic Press. New York, NY.

Soil Organic Matter. Schnitzer, M., and S.U. Kahn (eds.). 1978. Developments in Soil Science 8. Elsevier Scientific Publishing Co. Amsterdam, Holland.

"Soil organic matter and its dynamics." Jenkinson, D.S. 1988. pp. 564–607. In *Russell's Soil Conditions and Plant Growth* (A. Wild, ed.). John Wiley & Sons. New York, NY.

Soil Testing Laboratories

Most state land grant universities have soil testing laboratories. A number of commercial laboratories (such as Brookside and A&L Laboratories) also perform routine soil analyses. The ATTRA publication, *Alternative Soil Testing Laboratories,* is available on line (www. attra. org/attra-pub/soil-lab.html) as well as in print.

Publications

Soil Testing: Prospects for Improving Nutrient Recommendations. Havlin, J.L., et al. (eds). 1994. Soil Science Society of America. Madison, WI.

Soil Testing: Sampling Correlation, Calibration, and Interpretation. Brown, J.R., T.E. Bates, and M.L. Vitosh. 1987. Special Publication 21. Soil Science Society of America. Madison, WI.

Index

A

acidity, 27, 58, 148, 169–173, 187, 207
    aluminum solubility and, 28
    bacteria and, 15
    crop rotations and, 108
    manures, use of, 81
    sludges, use of, 72
    soil conditions, effect on, 35
aeration, 9, 22, 25, 41–44, 64, 208. *See also* compaction
    of compost, 112–113
    cover crops, effect of, 97
    decomposition rate and, 34
    earthworms and, 18
aggregates/aggregation, 9, 10, 26, 44, 206
    compaction and, 48–49, 125–126, 128, 136
    erosion and, 75
    rainfall, dispersal by, 47–48
    residues and, 68–69
    soil organisms and, 10, 17
    tillage and, 36, 121, 136–137, 140

alfalfa, 15–16, 86, 145, 172
    compaction and, 130–131
    as cover crop, 89–91
    in rotations, 37, 85, 102–105, 161, 164
algae, 9, 17
allelopathic effects, 95
aluminum, 9, 25, 28
ammonia, 83, 164
ammonium, 23–24, 30, 31, 70, 78, 82
animals, 9–10, 17–19. *See also* livestock farms

B

bacteria, 4, 9, 14–16, 20, 39, 44, 84, 99. *See also* nitrogen fixation
barley, 89, 102
berseem clover, 89–90
biological diversity, 74, 102
buckwheat, 89, 92, 95, 105
bulking materials, 112

C

cabbage, 52, 92, 103, 105

calcium, 9, 17, 23, 24, 58–59, 72, 75, 79, 108, 126, 148, 149, 168, 175, 187
carbon, 23. *See also* C:N ratio
    in compost, 110–112
    inorganic/organic, 11
carbon cycle, 28–29
carbon dioxide, 71, 115
    atmosphere, buildup in, 28–29
    as plant carbon source, 23
    respiration, production during, 9, 11, 13, 42
    soil aeration and, 9, 25
carrots, 84, 105
catch crops. *See* cover crops
cation exchange capacity (CEC), 24, 169–170, 175, 180–181
CEC. *See* cation exchange capacity
cereal rye. *See* rye
chelates, 25, 115, 149, 168
chisel tillage, 137–138, 140
clay, 9, 34, 49, 169, 173, 175
clay soils, 11, 24, 41–43, 64–65,

clay soils *continued*
86, 130, 155, 172
clovers, 15, 101, 102, 108, 161.
See also specific clovers
C:N ratio, 69–70, 91, 110–112
compaction, 4, 40, 44, 63, 136,
138, 207, 212, 213
cover crops and, 88, 97
load distribution, effect of,
129–130
plow layer and subsoil compaction, 126–133, 148
preventing/lessening, 6, 11,
25–26, 125–133
roots and, 19
tillage practices and, 141–144
traffic on soil, effect of, 129,
132–133
types of, 47–52
water range for, 54
compost, 59, 64–66, 68–69, 73,
105, 109–119, 147, 152,
211
advantages, 115–116
animals, composting of, 112
C:N ratio, 110–112
compaction and, 132
curing stage, 114
diseases suppressed by, 114,
115
making, 110–114
manures, 78, 109, 116–118,
151
moisture in, 110, 111, 113
pile size, 112
soil biodiversity, effect on, 74
starting materials, 110–112
turning pile, 112–113
use of, 115
controlled traffic, 129, 132–133
corn, 86
continuously grown, 120
cover crops mixed with, 145
grain, 57, 65–66, 72
manure application rates, 80,
82, 83
nitrogen for, 80, 82, 83, 159
residue from, 72
in rotations, 99–105

corn *continued*
silage, 37, 38, 65, 72, 80, 132
stalk rot, 148
sweet, 52, 97
cotton, 99
cover crops, 15–16, 65, 73, 87–
97, 126, 136, 145, 210,
212. *See also specific crops*
compaction and, 130–131
effects of, 88
for erosion reduction, 122
incorporation of, 131
intercrops, 95–96
mixtures of, 92
mulching of, 131, 142
mycorrhizal fungi and, 74
nutrients from, 75, 86, 148,
160, 163–164, 199
residue from, 64, 67–69
in rotations, 37–38, 103,
107–108
selection of, 88–89
soil biodiversity, effect on, 74
timing growth of, 92–95
types of, 89–92
for weed control, 74, 142
cowpeas, 90, 107
crimson clover, 89–90, 97, 107,
161
crop residue, 40, 64–68, 88, 152,
210–212
accumulations, effects of, 72–73
application rate, 72
availability, 102
burning, 67
characteristics, 68–73
C:N ratio in, 69–70
composting, 109
mulches, use as, 67–68
nutrients in, 56, 159
reduced tillage practices, 121,
126
removal of, 66–67
from rotations, 100–102
soil biodiversity, effect on, 74
crop rotations. *See* rotations
crown vetch, 91, 95
crust (soil), 47–49, 125–127,
136, 206

D
deep tillage, 128–129
disc harrow, 137–139
diseases, 4, 14, 17, 20, 67
compaction and, 74, 127
composts and, 74, 114, 115
earthworms and, 18
rotations and, 99, 102
tillage practices and, 143
diversion ditches, 122
drainage, 4, 41, 43, 75, 122,
148, 207
humus, effect of, 11
improving, 130, 141
topographical position and,
34–35
of topsoil, 22
drought stress, 53–54

E
E. coli, 84
earthworms, 4, 9–10, 14, 17–18,
204, 205, 207, 212
pores created by, 18, 27
rotations, effect of, 102
surface residues, effect of, 44
tillage practices, effect of, 146
erosion, 4, 6, 44–47
cover crops and, 95, 97
nitrates, loss of, 31
organic matter levels and, 65
reduction of, 119–123
rotations and, 101, 104, 107
soil organisms, effect of, 10,
19
tillage, effects of, 46–47, 211,
212
tilth, effects of, 26, 44
topsoil, loss of, 35, 45
water, caused by, 45, 75, 206,
207
wind, caused by, 45, 75, 123

F
fertilizers, commercial, 150–156.
See also nitrogen
application, 136, 147, 148,
155–156, 159, 181–183,
211

fertilizers, commercial, *continued*
  cost of, 145, 155
  nutrient flows, effect of, 55–
    56, 58–59, 158
  organic materials compared,
    152–154
  selecting correct, 154–155
field capacity, 43, 53
fresh residues. *See* organic matter,
  active
friable soil, 48–49, 207
frost tillage, 143
fungi, 4, 9–10, 16–17, 20, 26,
  31, 39, 44, 74, 99

G
grassed waterways, 122
grasses, 57, 89, 91–92
  compaction and, 130–131
  for erosion reduction, 38, 122
  as livestock food, 210–211
  manures, use of, 82, 83
  nitrogen from, 91, 160, 161,
    163
  no–till, use of, 141
  in rotations, 37–38, 93, 97,
    99, 101–105, 210
green manures. *See* cover crops
groundwater, 31, 104, 147

H
hairy vetch, 15, 89–90, 161–163.
  See also cover crops;
  legumes
  in cover crop mixtures, 92,
    145, 163
  in rotations, 104–105, 107
  weed problems from, 95
hay. *See* grasses
health, soil, 4–6, 63–76
hogs, 86
humus, 10–11, 22, 63–64, 109,
  169. *See also* carbon;
  organic matter
  cation exchange capacity,
    source of, 24
  root development and, 27
  rotations, effect of, 101
hydrologic cycle, 31

hyphae, 16, 44. *See also* fungi

I
insects, 9–10, 18, 20
  compaction and, 127
  cover crops and, 88
  earthworms and, 18
  mulches, control by, 67–68
  nutrient management and, 74,
    82, 148
  pores created by, 27, 44
  rotations and, 99, 102, 104,
    163
  tillage practices, effects of,
    143, 145
intercrops, 73, 95–96
iron, 23, 25, 169
irrigation, 82, 126, 174. *See also*
  runoff

L
leaching. *See* nitrate; runoff
legumes, 15, 75, 89–91, 148,
    151, 160–163, 199. *See*
  also nitrogen fixation;
  specific plants
  bacteria inoculation of, 89
  for erosion reduction, 122
  incorporation of, 138
  manures, use of, 82
  no–till for, 141
  root systems, 101
  in rotations, 37–38, 86, 92,
    99, 102–105, 210
lettuce, 65, 84, 107
lignin, 16, 68–70, 72, 80, 111–
  112
lime/limestone, 11, 59, 72, 75, 81,
  156, 168, 169, 171–172,
  211
livestock farms, 73, 131. *See also*
  manures
  composting of animals, 112
  groundwater pollution, 147
  nutrient cycling, 55–59, 150–
    151, 161–162, 164–165,
    213
  rotations, 103–105, 210–211
  tillage, 138, 143, 213

M
magnesium, 9, 17, 23, 24, 58–
    59, 72, 75, 79, 108, 147,
    149, 167, 168
manganese, 23, 25, 147, 149,
    168–169
manures, 38, 67, 70, 77–86, 166
  accumulations, effects of,
    72–73
  application, 65, 73, 82–84,
    105, 210–211
  chemical characteristics, 78–79
  compaction and, 132
  composting, 109, 116–118,
    151
  decomposition rate, 64,
    80–82
  green (*See* cover crops)
  handling systems, 78
  incorporation, 122, 138, 148,
    164
  nutrient content, 55–56, 75,
    77, 152, 160, 162, 167,
    199
  pollution from, 147
  residues, characteristics of,
    68–73
  soils, effects on, 58, 59, 74,
    80–82
  storage of, 78
  testing, 75, 79, 148, 159
micronutrients, 23–25, 147, 149,
    152
microorganisms, 15–17, 34, 44,
    64, 101. *See also specific*
  organisms
  C:N ratio, 69, 70
  in composting, 109–111
  root development and, 27
  tillage practices and, 37
millet, 105
moldboard plows, 37, 101–102,
    137–138, 143, 156. *See*
  also tillage
mucigel, 20
mulches, 65, 67–68, 73, 75, 122,
    126, 142
mycorrhizae (mycorrhizal fungi),
    16–17, 26, 44, 74, 92

N
nematodes, 17, 20, 74, 99, 102, 163
nitrate
in atmosphere, 30
availability, 24, 70, 75
in composts, 115
leaching, 31, 43, 55–56, 72, 82, 92, 104, 157–159, 163, 164
in leafy crops, 82
sod, use of, 163
nitrogen. *See also* C:N ratio
application rate, 72
in compost, 110–113, 115, 116
from cover crops, 88, 90–92, 95
deficiency, 69
denitrification, 31, 158–159, 164, 208
excess, 4, 148, 157
fertilizers, 30, 152–156, 162–163
immobilization, 70
inorganic, 30
management of, 75, 147, 149, 157–166
from manures, 72–73, 78–79, 82–83, 85
mineralization, 17, 23–24, 131, 135
nutrient flow patterns, 57–59
rotations and, 99, 102–105
soil tests for, 160, 162, 184–185, 211
tillage practices and, 143
use of, 122
nitrogen cycle, 29–31
nitrogen fixation, 15–16, 24, 25, 30, 59, 82, 89–91, 99, 158, 159
no–till, 18, 101, 103, 108, 120–122, 127–128, 131–132, 140–142, 145–146, 156, 211–212

O
oats, 89, 91–92, 97, 101–105, 107
onions, 57, 103, 105, 174

organic farms, 143, 151, 152, 154, 212–213
organic matter
active, 9, 10, 39, 64, 101, 116, 149
adding, 38, 122, 126
amount in soil, 33–40, 44, 148
application rates, 72–73
C:N ratio in, 69–70
distribution in soil, 38–39
fertilizers *vs.*, 152–154
importance of, 21–33
levels (*See* organic matter levels)
living, 9–10, 13–20, 39–40, 103 (*See also* soil organisms)
management of, 5–6, 63–73, 104, 150
parts of, 9–11
use of, 65–68
organic matter levels, 6, 33–38, 64–65
maintenance of, 122
nutrient availability, 149
rotations, effect of, 100–102
oxidation, 11
oxygen, 4, 9, 11, 13, 23, 25, 75, 112

P
peas, 15–16, 86, 108
penetrometer, 133, 204, 206
peppers, 105, 145
pH. *See* acidity
phosphorous, 23
cation exchange capacity and, 24–25
in composts, 115
excess, 4, 72, 157–158, 166–167
fertilizers, 152–156, 162–163
management of, 147–149, 157–166
in manures, 78–85
mineralization, 15–17, 23–24, 149
nutrient flow patterns, 57–59
reducing losses of, 162–164
soil tests for, 108, 185–187, 211

photosynthesis, 11
physical condition of soil, 41–54, 63, 115, 150
Phytophthora, 127
plant tissue tests, 182–183
plastic (soil), 48–49
plastic limit, 48–49
plow layer, 47–49, 126–133
pores, 9, 26–27, 34, 40–44, 52, 81
potassium, 9, 23, 167
cation exchange capacity and, 24
in composts, 115
excess, 166, 167
fertilizers, 152–156
management of, 147–149
in manures, 78–79, 81, 83
nutrient flow patterns, 57–59, 75
in sewage sludge, 72
soil tests for, 108, 211
potatoes, 65, 84, 103, 105, 132
protozoa, 9, 17, 20

R
rainfall, 4, 34, 82, 121, 125, 126, 140, 208. *See also* runoff; water
rape, 89, 92, 97
red clover, 86, 89, 91, 92, 105
respiration, 11, 13
rhizobia, 15, 89. *See also* legumes; nitrogen fixation
ridge tillage, 140–141
roots, 19–20, 44, 75, 148
cells, respiration of, 9, 25
compaction and, 52–53, 126, 130–131
in degraded soils, 41
diseases, 74, 114, 115
for erosion reduction, 122
evaluation of, 206–207
fungi and, 16
left after harvest, 66–67
nutrients and, 70, 150
old channels of, 27
rotations and, 100–103
of sods/grasses, 35, 38

roots *continued*
    soil organisms, interactions with, 10, 20, 27
    stimulation of, 27
rotations, 99–108, 145, 160, 161, 163, 210, 212
    biodiversity of soil, effect on, 74
    compaction and, 126, 130–131
    economics of, 102
    erosion, effect on, 124
    examples, 103–105
    farm labor and, 102
    general principles of, 102–103
    incorporation of, 131, 138
    manure use and, 85
    mulching of, 131
    organic matter levels, effect on, 37–38, 40, 63, 73, 75, 100–102
    rooting periods, 102
    species richness for, 102
runoff, 4, 45, 58–59, 122, 148, 206. *See also* nitrate, leaching
    cover crops/rotations and, 75, 87
    manure incorporation, 82–84
    surface crusts, effect of, 47
    tillage practices and, 120
rye (winter, cereal), 74, 88, 89, 91, 92, 94, 104–105, 107, 130, 131, 145, 163
ryegrass (annual), 92, 95, 97

S
safflower, 103
saline seep, 105
saline soils, 6–7, 173–175
salt damage, 82, 84, 105, 187, 211
sand, 9, 34, 41, 64–65
sandy soil, 11, 28, 34, 41–43, 49, 65, 86, 125, 168, 169
sewage sludge, 59, 70–73, 132, 147
snap beans, 52
sod. *See also* grasses; legumes
    continuous growth of, 100

sod *continued*
    erosion, effect on, 120
    in rotations, 99–102, 104, 105, 120
    tillage, 131
sodic (alkali) soils, 168, 173–175
sodium. *See* saline soils; salt damage; sodic (alkali) soils
soil aggregates. *See* aggregates/ aggregation
soil color, 28, 205, 207
soil consistency, 48–49
soil degradation, 6–7, 41, 45, 46
soil hardness, 204–206
soil loss tolerance, 119–120
soil organisms, 4, 69, 102, 205. *See also* microorganisms
    beneficial effects of, 25
    classification of, 13–14
    cultivation, effect of, 40
    size of, 14
soil quality, 4–6, 63–76
soil solution (water), 9
soil strength, 52
soil structure, 4, 9–10, 44. *See also* aggregates/aggregation
soil testing, 60, 108, 156, 159–161, 177–199, 205, 208, 211
    accuracy of, 178
    adjusting recommendations, 198–199
    cation exchange capacity, 170
    examples, 188–197
    interpreting results, 187
    manures (*See* manures)
    nitrogen, 184–185
    for organic matter, 187
    phosphorous, 185–187
    samples for, 177–178
    variations in, 179–184
soil texture, 34, 49, 52, 64
sorghum, 65–66, 103, 132
soybeans, 15, 86, 90, 168. *See also* legumes
    residues from, 65, 68–69, 132
    in rotations, 37, 99, 104, 107, 145
spiders, 18

squash, 105, 107
strip cropping, 103, 122–123
strip tillage, 97
subsoils, 35, 39, 103
    compaction, 47, 50–52, 126–133, 207
    deep tillage, 128–129
subterranean clover, 89–90
sudangrass, 92, 97, 107, 131
sulfur, 23, 24, 79, 83, 147, 149, 168
sunflower, 103, 105
surface crusting. *See* crust (soil)
surface sealing, 125–126
sustainable agriculture, 5
sweet clover, 89, 91, 162

T
terraces, 45, 123
tillage, 36–37, 39, 48–52, 73, 97. *See also* moldboard plows; no–till
    compaction and, 127–129
    conventional, 137–140, 143
    erosion and, 120
    fertilizer/amendment incorporation, 156
    frost tillage, 143
    organic matter levels and, 65
    reduced, 121, 126, 135–146, 160, 164, 211–213
    rotation of systems, 143
    selecting best practice, 141–143, 145
    soil check prior to tilling, 50
    soil degradation and, 46–47
    timing of field operations, 143–144
tilth, 5, 9, 22, 25–27, 44, 63, 122, 148, 163, 173, 205–207
timothy, 101
tomatoes, 57, 68, 105, 107, 145, 148, 175
topographical position, 35, 46–47
topsoil, 15, 35, 39, 65
    erosion and, 119–120, 122
    organic matter content, 21–22
    properties of, 22
    rotations, effect of, 100–101

traffic, controlled, 129, 132–133

V, W, Z

vetch, 86, 97. *See also* crown
vetch; hairy vetch
water, 9. *See also* groundwater;
rainfall; runoff
cycle, 31
optimum range for plant
growth, 53–54
organisms living in, 17
in soil pores, 42–44
water infiltration, 44–45, 150, 207
compaction, effect of, 47
composts and, 117

water infiltration *continued*
cover/rotation crops, use of,
103, 131
earthworms and, 18
organic matter levels and, 65
salts, movement of, 105
soil structure and, 4, 10, 31
water storage, 4, 22, 36, 65, 207
water supply, 75, 148
cover crops and, 89, 95
manure composting, effect of,
116
mulches and, 67
weed control, 4, 67, 74, 87, 163
compaction and, 127, 131

weed control *continued*
cover crops and, 88–90, 92
nutrient management and, 148
rotations and, 99, 103–105,
108
tillage practices, 135–136,
138, 142, 143
wheat, 37, 86, 89, 97, 101–102,
104–105, 107, 120
white clover, 89, 91
wilting point, 43, 53–54
winter rye. *See* rye (winter, cereal)
zinc, 23, 25, 79, 83, 147, 149, 168
zone tillage, 129, 131, 132, 138–
142